CULTURAL LANDS
POST-SOCIALIST

Re-materialising Cultural Geography

Dr Mark Boyle, Department of Geography, University of Strathclyde, UK and
Professor Donald Mitchell, Maxwell School, Syracuse University, USA

Nearly 25 years have elapsed since Peter Jackson's seminal call to integrate cultural geography back into the heart of social geography. During this time, a wealth of research has been published which has improved our understanding of how culture both plays a part in, and in turn, is shaped by social relations based on class, gender, race, ethnicity, nationality, disability, age, sexuality and so on. In spite of the achievements of this mountain of scholarship, the task of grounding culture in its proper social contexts remains in its infancy. This series therefore seeks to promote the continued significance of exploring the dialectical relations which exist between culture, social relations and space and place. Its overall aim is to make a contribution to the consolidation, development and promotion of the ongoing project of re-materialising cultural geography.

Other titles in the series

Towards Safe City Centres?
Remaking the Spaces of an Old-Industrial City
Gesa Helms
ISBN 978-0-7546-4804-8

Fear: Critical Geopolitics and Everyday Life
Edited by Rachel Pain and Susan J. Smith
ISBN 978-0-7546-4966-3

In the Shadows of the Tropics
Climate, Race and Biopower in Nineteenth Century Ceylon
James S. Duncan
ISBN 978-0-7546-7226-5

Geographies of Muslim Identities
Diaspora, Gender and Belonging
Edited by Cara Aitchison, Peter Hopkins and Mei-Po Kwan
ISBN 978-0-7546-4888-8

Presenting America's World
Strategies of Innocence in National Geographic Magazine, 1888-1945
Tamar Y. Rothenberg
ISBN 978-0-7546-4510-8

Cultural Landscapes of Post-Socialist Cities
Representation of Powers and Needs

MARIUSZ CZEPCZYŃSKI
University of Gdańsk, Poland

Routledge
Taylor & Francis Group

LONDON AND NEW YORK

First published 2008 by Ashgate Publishing

2 Park Square, Milton Park, Abingdon, Oxon OX14 4RN
711 Third Avenue, New York, NY 10017, USA

Routledge is an imprint of the Taylor & Francis Group, an informa business

First issued in paperback 2016

British Library Cataloguing in Publication Data
Czepczyński, Mariusz
    Cultural landscapes of post-socialist cities :
    representation of powers and needs. - (Re-materialising
    cultural geography)
    1. City planning - Europe, Eastern 2. Landscape protection
    - Europe, Eastern 3. Landscape - Symbolic aspects - Europe,
    Eastern 4. Post-communism
    I. Title
    307.1'216'0947

Library of Congress Cataloging-in-Publication Data
Czepczyński, Mariusz.
    Cultural landscapes of post-socialist cities : representation of powers and needs / by
Mariusz Czepczyński.
    p. cm. -- (Re-materialising cultural geography)
    Includes index.
    ISBN 978-0-7546-7022-3 (alk. paper)
    1. Human geography--Europe. 2. Urban geography--Europe. 3. Urban renewal--Europe.
4. Public spaces--Europe. 5. Landscape changes--Europe. 6. Post-communism--Europe. 7.
Czechoslovakia--History--Velvet Revolution, 1989. I. Title.

    GF540.C94 2008
    304.2094--dc22

                                                                            2007051239

ISBN 13: 978-0-7546-7022-3 (hbk)
ISBN 13: 978-1-138-25427-5 (pbk)

Contents

List of Figures

List of Boxes

Acknowledgments

This project emerges from a long-standing interest in visual aspects of cultures and their representations, but was completed only thanks to discussions, interpretations, sometimes argues and encouragements from number of people. As I worked on this project I came to realize how broad was the scope of the problem I set out to address. The process of improvements, maturity and proceeding was long and complex. This book has been written, rewritten, corrected, and replenished a number of times over the last three years. There are many people and institutions that helped me to realize the project and eventually write the book. The realization of the project would not be possible without the support and permanent assistance of my colleges from the Department of Economic Geography of the University of Gdańsk, under the guidance of Iwona Sagan. Field studies were the most crucial part of my research, and my explorations of Central European cities were only possible due to the financial support of the Department of Economic Geography, dean of the Faculty of Biology, Geography and Oceanography and the Rector of the University of Gdańsk. Implementation of the project was also possible due to a grant, received from the Polish Committee of Scientific Research (Komitet Badań Naukowych), now Ministry of Science and Higher Education (Ministerstwo Nauki i Szkolnictwa Wyższego). Initial inspiration and stimulation for this book came from seminars at Queen Mary University of London and Hull University in 2003. It would be impossible to thank everyone who has influenced my ideas in conversations, consultations and meandering disputes, but will start with Roger Lee from Queen Mary University of London, Alison Stenning, University of Newcastle, Bolesław Domański, Jagiellonian University in Kraków and Vera Denzer, University of Leipzig. Publications and advice received from Anita Wolaniuk and Krystyna Rembowska from the University of Łódź and Jonathan Oldfield, University of Birmingham. Several long and fertile disputes with my friends Piotr Szafranowicz and Joanna Małuj from Gdańsk, Philipp Rode from Vienna, Matt Fergusson and Dirk Welsch-Lehmann from Berlin help to synthesize and categorize many of the landscape transformation processes. Special thanks are also dedicated to all the people I have talked during my field trips around Central Europe. All those questions, conversations and consultation in trains, taxis, cafés, bars, buses, universities, baths, streets, parks, shops and museums were most helpful in understanding and interpreting cultural landscapes.

I would also like to thank the Ashgate team – Val Rose, Gemma Lowle and others, for belief, reliance and patience. I am especially appreciative of the support and patience of my parents, Antoni and Teresa, my brothers Jarek and Kazik, and my sister Bożena. My best friends in Sopot, Gdańsk, Gdynia, Berlin, London, Bonn and Vallenciennes backed me immensely during the challenging time of completion of the book.

To my parents Antoni and Teresa, my primary landscape interpreters

Introduction

Humans' spatial position has long been connected with human experience, existence and life. This interest has sprung from the crucial need for co-ordinating important environmental relations, and for implementing meaning and order into the world of experiences and emotions. One's relation to surrounding objects can be cognitive or affective, always aiming to create a certain level of equilibration between man and environment. The land in which we live both shapes us and is shaped by us: physically, by means of cultivation and building, and imaginatively, by projecting our aspirations and fantasies of wealth, refuge, well-being and awe (Norberg-Schulz 1971, Robertson and Richards 2003). The everyday space of modern man does not comprise an entire universe, as was the case in early civilizations. Space has been fragmented into various specialized components, used for orientation and adaptation. Among many diverse cognitive spaces, several psychological components can be distinguished, like the immediate space of perception and fairly stable spatial schemas. The schemas are the result of the personal accommodation of various signs along with their assimilation. The representations consist of universal elementary structures – archetypes, as well as socially or culturally conditioned configurations and personal idiosyncrasies. Together, they create a picture of the surrounding space in the appearance of a particular form of landscape, usually as an established tri- or quarto-dimension system of relations between the meaningful objects (Piaget, 1968). As Bender (2001, 4) rightly put it:

> It is through our experience and understanding that we engage in the materiality of the worlds. These encourages are subjective, predicated on our being in and learning how to go on in the world. The process by which we make landscapes is never pre-ordained because our perceptions and reactions, though they are spatially and historically specific, are unpredictable, contradictory, full of small resistances and renegotiations. We make time and place, just as we are made by them.

Cities act as a totalizing and almost mystic gauge of socio-economic and political strategies; urban life allows what has been excluded from it by the urbanistic plan to increase even further. The language of power is 'urbanized'. But the city is subjected to contradictory movements that offset each other and interact outside the purview of the panoptic power. City becomes the dominant theme of political epic but it is no longer a theatre for programmed, controlled operations. Beneath the discourse ideologizing it, there is a proliferation of tricks and fusions of power that are devoid of legible identity, that lack any perceptible access and that are without rational clarity (Certeau 1985, 127–8). City has both structured and structuring qualities. Landscape structures people's perception, interactions, and sense of well-being or despair, belonging or alienation. The structuring qualities are most clearly felt (and most visible) in the built environment, or landscape, where people can erect homes, react to architectural forms, and create – or destroy – landmarks of individual or collective meaning. Space also structures metaphorically, stimulates both memory and desire. Cultural landscape can be seen as a major structuring medium (Zukin 1993, 268). According to Ratzel, cities are 'the best place to study people because life is blended, compressed, and accelerated' in cities, and they bring out the greatest, best, most typical aspects of people (Wanklyn 1961, 36).

Each set of social and economic structures coexists with a diversity of aesthetic forms, material practices, and political institutions. Starting with the basis of the social world, many studies aim to explore interrelations of social structure, especially institutions of power, class, and social reproduction, or the forms that represent, transmit, and transform institutionally embedded power relations. There is a widespread tendency to integrate space and time into the description and explanation of social patterns. There is also a focus on mutual effects of economies, politics and culture in the restructuring processes. Changes in the economic system are treated equally with changes in the cultural context of social and economic behaviour (Zukin 1993, 21). Cultural landscape, as an integral part of the studies of powers, must 'embrace material practices as well as aesthetic forms, underlining the convergence between economic structure and cultural project, representing the experience of all social classes'. Urban landscape has been seen for ages as being among the main composers as well as transmitters of culture. Architecture is the bodily expression of the ways of thinking, the experience, and the hierarchies of values and culture of each of the group as well as of each individual. This 'true knowledge', culturally conditioned *epistemé* is always boldly visible through the forms of urban space and its organization. Architecture is one of the main representing languages of modern society that signifies the spiritual dimension of investors, architects and users (Czepczyński 2005a). The buildings are central to understanding the landscape in that they frame and embody economic, social and cultural processes. The aesthetic form is never neutral – the power is written into the landscape through the medium of design. There are two, generally speaking, approaches to understand landscapes, often seen as contrary or opposing. Some of the researches, mostly ecologists and urbanists, understand landscape as an entity, while anthropologists and historians see landscape as a relationship. I will try in this book to merge both of the approaches, and interpret cultural landscape as an *entity reflecting relationships*.

Our cognitive maps, aesthetic forms and ideologies reflect the multiple shifts and contrasting patterns of growth and decline that shape the landscapes. 'Landscape is the major cultural product of our times', declared Zukin (1993, 22). In history of painting, landscape includes both real scenes and the perspective from which we view them. Cultural landscape can be interpreted as an announcement/revelation/exposure of information about local societies. Sometimes landscapes, decoded of hidden meanings, expose more information then we are prepared to hear. Interpretation of cultural landscape reveals many layers of information which can be used according to current social and economic circumstances. Built or anthropogenic landscape always mirrors culture of its creators. The systems of values, preferences, beliefs, fears and 'truths' are expressed in landscape features and settings. This 'true knowledge', culturally conditioned epistemé is always boldly visible through the forms of urban space and its organization. Architecture is one of the main representing languages of modern society, which signify the spiritual dimension of the investors, architects and users

(Czepczyński 2005a). Buildings are central to understanding the landscape in that they frame and embody economic, social and cultural processes. The aesthetic form is never neutral – the power is written into the landscape through the medium of design.

Contemporary processes of cultural landscape transformation are integral elements of the geographical divisions of space. Core landscape transformations usually follow chief social evolutions or revolutions. One of the recent major political transformations had been started in Central Europe in 1989 with the collapse of communist regimes. Cultural landscape of Central Europe carry many features that correspond to communist powers, structures and procedures, represented by buildings and urban settings. One of the most visible and therefore most irritating post-socialist heritages include iconic landscapes. In consequence contested anti-communist values and ideologies were followed by anti-communist landscape purges and cleansing, assisted by implementation of new concepts and ideas. The concept of 'Central Europe' has been undergoing renaissance, initiated by popular Czech writer Milan Kundera in the mid-1980s. The concept was supposed to be a particular 'anthodium' to the Soviet bloc ethos, and was further developed in opposition in most popular in the Western naming of everything east of Elbe as Eastern Europe by some prominent historians like Norman Davis or Timothy Garton Ash. Since for many of the inhabitants of the region, including the author, the term 'Eastern Europe' is not geographically relevant, and more importantly relates us to the Soviet empire and Russia, we could rather call ourselves 'Central' Europeans.

There is an 18-year-old tradition of analyzing, synthesizing, interpreting, forecasting, modelling and understanding post-socialist reality. The 'post-socialist difference' (Hörschelmann 2005) has been studied at various regional levels and by diverse specialists, including economists, sociologists, historians, geographers, architects, anthropologists, and many others, who differently emphasized the past and ongoing processes and structures. Hundreds of books have been written, including dozens of monographs. The diversity of approaches can be illustrated by such different publications on post-communism as Holmes (1997) and Staniszkis (2005). There have also been thousands of articles published, both in scientific periodicals and in the press and magazines; most of the publications have appeared in the national languages of Central Europe, including Polish, German, Hungarian, Czech, Russian and many others. In the large body of literature produced over last twenty years on post-communist transformation, studies investigating changes in urban structures and landscapes have been fairly rare (see Stanilov 2007). Public attention has been on matters related to the economic and political transformations considered central for the advancement of social reforms.

The notion of the term 'post-communism' or 'post-socialism' has undergone countless debates and generally speaking carries two perspectives. For most researchers from Central and Eastern Europe, post-communism is directed and related to the past and means the burdensome relations with the communist regimes or pejorative social, cultural, economic inheritance. For the majority of Western analysts, post-communism means hardly anything more than the general social, political, economic and cultural reality in Central Europe after communism era. The two opposing constructions of post-communism are additionally supplemented by discourse between the two groups. While researchers from the East are often accused of lack of objectivism, situated knowledge and no relation to the most current Western (English-language) discussion on post-socialism; Western researchers frequently display paternalism studying 'Eastern wildness', applying their *besserwisser* and 'exotic studies' attitudes (see for example Domański 2004a 2004b, Stenning 2005). Central Europe is of interest because studying it yields insight into how nations of the former Soviet Bloc are interpreting their years of Communist rule and how commemorative traditions have respond to changes of political regime. This trait is shared with the many nations that experienced political upheaval and revolution in the 19th and 20th centuries (Argenbright 1999).

There is an often-raised question of when 'post-communism' ends and for how long Central Europe is going to be labelled as 'post-communist'. Despite some authorities' declarations through social research that 'post-socialism is over', a lot of socialism can be seen on the streets of Central European cities. In some aspects, societies between the Elbe and Bug, and the Baltic and Black Sea are not socialist any more, but in some other places the post-communist burden is still clearly seen in many aspects of social, economic and political life. Krzysztof Król (2007, 12), a former Polish anti-communist leader, said that 'leaving communism will take us another 22 years. Moses walked its people through the desert for 40 years, until all generation grew in servitude will die. We have walked for 18 years by now'. The growing generation of 40-year-olds seems to play a more and more important role in the cultural and economic life of the post-socialist countries. We still remember the last years of communism; we have consciously experienced the transformation of 1990s. We all carry the stock of experience and memories, although are usually too young to have been too deeply involved in communist discourse and interactions, but the same time have had time for reflection. We will probably never create any comprehensive 'schools', but will look and interpret on our own way.

Despite its historical and geographical differences, the region shares, a number of common characteristics relevant to the management of its heritage. Its geopolitical position between German–Habsburg and Russian–Slavonic realms and subsequent history has created a social and ethnic spatial complexity. The survival of dynastic states associated with the relatively late development of nationalism and the nation state has left ethnic enclaves and exclaves, national minorities and irredentist possibilities. There was an abrupt post-Second World War suppression of nationalism and ethnic regionalism within the Soviet political hegemony, and finally a sudden economic and political transition in the 1990s, consequent upon the collapse of that hegemony and the rise of new national self-awareness and new international orientations (Ashworth and Tunbridge 1999, 105–106). 'Citizens of countries with 50 years' experience of the operation of Marxist historical determinism do not need convincing that among these contemporary needs, and consequent roles of heritage, is the political legitimation of governments and governing ideologies. It is, perhaps, less obvious to argue that such need for legitimacy is not confined to totalitarian regimes but is just as necessary, although more multifaceted, in pluralist democracies with liberal free market economic systems' (Ashworth and Tunbridge 1999, 155).

According to Duncan (1990, 18) the job of the cultural geographer is 'to show how the locals' accounts are constituted within a system of signification, connected to the other elements within the cultural system produced within a social order. The second line of inquiry is non-local accounts. (...) The juxtaposition of insiders' and outsiders' readings can help to defamiliarize the relationship between landscapes, dominant ideologies, and political and social practices. It can illuminate the way dominant ideologies which are communicated through the medium of landscape reproduce social and political practices.' Foucault suggests that power is a dispersed set of micro-practices, many of which operate through the normalizing gaze of surveillance regimes (Dovey 1999). The thousands of landscape practices are implemented by hundreds of large, medium and small-scale landscape lords and all together fill the contemporary urban landscape jigsaw. Although the post-socialist landscape transition is a unique and distinctive process, the procedures or micro-practices are similar if not the very same as at any cultural landscape transformations.

Recognition of the external world, environment or space is complex and multi-aspect process, consisting of perception, but also recognition, comparison, classification, and valuation. This process often modifies and conditions our behaviour (see Massey 2006). The challenge to decode, interpret and explain seems to be very appealing, but also demanding. Various understandings of landscape, together with multiple methodologies borrowed from linguistics and other social sciences require both interdisciplinary knowledge and submission. Explaining and interpreting cultural landscape is always a personal experience. Continuously choosing facts, adding texts to features is always personally conditioned, and based on personal history, experiences, beliefs. Hubert Beuve-Méry, the establisher of the French daily *Le Monde*, when asked if journalism can be objective, apparently said 'of course not'. The interpreters of the world should be selflessly/disinterestedly subjective, in opposition to the objective mercenary/interestedness/selfishness. The question of objectivity and subjectivity in social sciences will be further mentioned and evolve. One of the main research problems in cultural studies, including cultural geography, is the question of engagement into the research field. On one hand, someone from the 'within' might be less objective and personally conditioned by his or her history and experiences. On the other hand, a locally anchored researcher might have quite a broad knowledge of regional circumstances and conditions.

Synthetic descriptions and interpretations are among the oldest geographical research studies, creating the base for geography, as plainly translated from Greek: 'describing the Earth'. The shape of cultural landscape can be explained as a form of conceptualization of the world, representing the society that made the landscape. New, post-Berkley cultural geography is based on culture as a holistic, complex and dynamic structure, represented by many various everyday objects, features and places. Every landscape perceived by humans carries a smaller or larger dose of cultural meaning, coded in forms and signs. In consequence, cultural landscape research is based on intuition, empathy, reflections, direct and indirect experiences of places, buildings and events, contextualization, employing analogies to understand/ decode the landscape signs/language. The significance of cultural landscape studies, as synthesis of the visual and the hidden, the intention and the outcome, relies on relations between different components and users of the landscape. I try to combine two central aspects: we focus both on the material arrangement of the places and the way individuals live and realize the different places with the variety of meanings attached to them.

The synthesis, notwithstanding aspirations and abilities, will be somehow deformed or filtered through my stock of experiences, or, generally speaking, my culture. There is a long debate in social science between advocates of internal and external points of research. Is it better to have more objective view of the outsider, or rather, should regionally embedded studies be done by local researchers who might have more intuition and more familiarity with the object of study, but who are involved, on many layers, in the processes they analyze? I am far from giving the answer to that question. The dilemma has been discussed by many (see Stenning 2006). My social, economic, cultural and historical position somehow shapes the choice of my research methods and perception.

This book aims to present the general outlook of post-socialist landscape conversions and attempts to structuralize practices and procedures of landscape transformations in Central Europe. The alteration of cultural landscape is synthesized by inductive reasoning, in which the premises of arguments are believed to support the conclusion. The book will ascribe properties and relations of cultural landscape to types derived from my experiences and interpretations to formulate more general trends and inclinations based on explanations of recurring phenomenal patterns. Cultural landscape explanations are generally derived from induction or inductive reasoning, based on the belief that premises of an argument support the conclusion but do not ensure it. Induction is used to ascribe properties or relations to types based on signs, sometimes on one or a small number of experiences to formulate more general trends founded on limited observations of recurring phenomenal patterns. Dozens of minor or major landscape features and their structural, functional or sensual characteristic will support broader conclusions and will help to synthesize the ongoing processes.

The classification and organization of landscape transformation processes is based on my comparisons, observations, and interpretations of wide selection of source collected during many trips through Central European countries in 2005, 2006 and 2007. Multiple sources correspond with different features of cultural landscape. Transformation of form is the easiest to trace, and was mainly based on observation and photographic documentations. Correspondingly, functional alternations are generally straightforward, connected with form, usually advertised or expressed on the structure. Decoding the significance of cultural landscape is always the most difficult and controversial part of the process. Tracing many social constructions and discourses, often in unpublished or dispersed information, was the most complicated task of the project. Ephemerid sources are often unrecognized by conservative researchers, and are also difficult to verify. Newspaper articles, web pages, interviews and conversations, promotional materials and leaflets, together with some publications focused on meaning of landscape were most important to decode significance of cultural landscapes. There are eight official languages used in the region, and what make the synthesis even more difficult, since lots of local sources can be only read in local languages. My knowledge of Polish, together with limited understanding of German, Czech, Slovak and very poor Bulgarian gave me access to some local sources, while in Hungary and Romania I had to base my research on translated, secondary information.

Changes of meanings, forms and functions of the landscape are exemplified by a set of subjectively chosen examples from various countries around Central Europe, understood as former European socialist countries incorporated into the European Union. The investigation was based on personal observation and exploration of cultural landscapes of Poland, East Germany or former German Democratic Republic, Czech Republic, Slovakia, Hungary, Romania and Bulgaria. The study does not include illustrations from former Yugoslavia, Albania and former Soviet Union. Many of the described landscape practices, like reinterpretations of communist icons, gated communities or revival of religious and nationalistic landscapes, are present in most of the post-socialist countries in Europe, including Russia, Ukraine, Estonia or Serbia. However, the transformation process in the former socialist federations of Yugoslavia and Soviet Union diverge from the rest of the former Eastern Bloc. Central Europe, despite multiple analogous experiences and processes, is significantly internally differentiated, not only between countries, but also between regions and cities. Comparable processes of post-socialist transformations are differently experienced in every of the analyzed cities. Even so the research is mainly focused on equivalent similarities, while local differentiations will be used to exemplify chosen processes. The project, following inductive reasoning, does not focus on few chosen cities – cases, although a numbers of cases, referring to many different processes are present in compact boxes. The boxes and a selection of my pictures provide more detailed exemplification of the landscape transformations.

The work on post-socialist landscapes can be seen as particular inventory of the process. Main research questions include how has cultural landscape been changed after 1989, and how much do those changes say about us? The central hypothesis of the book is how social, economic and cultural transformations of post-socialist societies of Central Europe are reflected in cultural landscape transformations. Chapters 4 and 5 are about negotiating boundaries and reconstructing landscapes. They are about landscapes as dynamic socio-geographic entities that are constantly being reshaped by political, economic and social forces as well as shifts in moral and value systems. These changes are both created of, and created by, people as they negotiate their place in the world. People attach themselves to particular landscapes, places and locales, but those places and people's sense of attachment is always in flux (Aziz 2001). Memory, oblivion and past experiences, as well as government interventions facilitate transformations of old socialist features, as presented in Chapter 4, while global penetrations of capital, as well as socials and cultural alternation, together with new hopes and expectations feature new elements of cultural landscapes, as shown in Chapter 5. All of those transformations together have major effects on people's sense of self, of place, and most importantly in this analysis, people's sense of landscape.

I do not aspire to present a comprehensive and the only interpretation of post-socialist landscape alternations. I would rather be subjective but as selfless as possible, in point of view and explanation. The theme of the project, 'post-socialist landscape', can be identified only in historical and social context. First, it is necessary to outline what is/was a socialist landscape and what features and processes are/were characteristic for the communist landscaping? The only possible way to delimit what was socialist and what was not is the comparison or contextualization with the outside word. Finding differences between Western European cities and the ones left behind the Iron Curtain help to understand, classify, and label socialist landscapes. The broader context seems to be the only method to differentiate one from another, and eventually to label the socialist and post-socialist landscapes. My *flânerie* around Vienna, Hamburg, London, Paris, New York and Barcelona helped me to contextualize my home/regional experience of Eastern European landscapes.

The book tries to answer the question of how ideological and political discourses and disputes are reflected in architecture, spatial organization and place-making in contemporary Central and post-socialist Europe. I find the relationship between political and economic powers and support and suggested way of creating cultural landscapes most intriguing and interesting: how ideologies inspire and influence contemporary architecture. The relation between investor as 'the power' and landscape is based on the assumption that the investor's ideology includes image of a preferred or perfect surrounding world, in which he or she would like to live and be. Most of that imagination is based on the investor's, architect's or user's personal history. Exploring and trying to understand a post-socialist landscape of traumatic memories and features, local and national pride and happiness, dynamic transformations, almost day-to-day changes is a great challenge. I try to read and decode the texts written into post-socialist landscape of Central Europe and sometimes attach a text to the existing feature.

The book is structured in six chapters. Two opening ones are focused on theoretical aspects of cultural landscape research and tradition. Understandings, schools, methodologies and interpretations of landscapes form the first chapter. Chapter 2 comprises representation, power and history in cultural landscape studies. Chapter 3 is historical and presents the main socialist era landscaping practices. Chapters 4 and 5 are the core of the book, and introduce two main themes: reinterpretation and re-contextualization of the old communist landscape icons and text, and introduction and interaction with global, contemporary process and landscape practices. The last chapter attempts to systematize and categorize the ongoing landscape processes. The book you are about to read was designed to give a general and holistic view of the transformations of post-socialist societies, reflected in forms, functions and meanings of cultural landscapes. The synthetic and inductive approaches might help to encompass diverse development paths in disparate, but similar societies.

There are many weaknesses of this book. It might be seen as too general, but a certain generalization was necessary to cover the very broadly defined field of the study. The language is often deliberately colloquial, used to better describe the language of the analyzed times and societies. The research has been largely based on primary sources, including private and official interpretations of landscape and much less on post-socialist, predominantly British, publications. The ongoing scientific discourse was most useful in general landscape and transformation analysis, while published cultural landscape studies have been quite rare. After a few years of working on this book, it does not look complete, but I do not believe it would be ever finally completed. Everlasting and ever-changing process of representation of powers and desires in landscape features, enriched by countless interpretations, make this book enduringly open for re-contextualizations and reinterpretations.

Chapter 1

Geographical Studies of Cultural Landscape

The notion of landscape, especially with cultural implication, has recently come to have an expanding resonance. Its material, economic and representative values are being increasingly employed in social, economic and political spheres, as well as in everyday life. Landscape, and in particular cultural landscape, has become one of the widespread meta-words of post-modern spatial and cultural discourses, incorporating dozens of possible interpretations and areas of interest. It seems that the flexible and elastic concept of landscape meets the growing demand for holistic and synthetic interpretations in many research fields. More and more often 'landscape' not only denotes the traditional meaning of physical surroundings, but also refers to the 'ensemble of material and social practices and their symbolic representation' (Zukin 1993, 16). Cultural landscape then becomes a social product, which embodies many representations of various local and global powers, as well as everyday practices and interpretations of more or less common users. Landscape then is much more than we can see; visual features are being facilitated by their functions and additionally enhanced by their socially constructed significance. Urban scenery becomes not only buildings, spaces and places, but also 'expressions of cultural values, social behaviour, and individual actions worked upon particular locations over a span of time' (Meining 1979, 6).

The term 'landscape' can be interpreted in a number of ways, which, though not mutually exclusive, differ in emphasis. Probably the most popular, visual meaning of the term 'landscape' is based on morphology, and involves the examination of the visible phenomena of the studied region or place. Many authors suggest (see Naveh and Lieberman 1984) that the earliest written word meaning landscape was cited in the Book of Psalms. This word – *noff* in Hebrew – concerns mainly the perception of a landscape, giving importance to the visual aspect, and it seems that after three thousand years the very same and basic meaning of *noff* is still based on experience of the visual appearance of land.

Contemporarily, the old biblical word has been replaced by the concept of landscape, which comes from the Dutch word *landschap*. It was borrowed as a painters' term during the 16th century, when Dutch artists were on the verge of becoming masters of the landscape genre. *Landschap* had earlier meant not much more then a 'region, a tract of land' but had acquired the artistic sense, which it brought over into English and other European languages, as 'a picture depicting scenery on land'. Soon after, the art of creating desired landscapes, known as landscaping, had developed in Italy and other European countries. 19th century faced the beginning of modern sciences and methodologies, and in the end of the century the concept of landscape had been incorporated into geographical research. By the end of 20th century, landscape, without losing its prior meaning, has been incorporated into architecture and anthropology. In consequence, since the 1980s landscape began to be regarded not only as an entity existing in the physical world but as a metaphor, text, image or scene as well (see Birks et al. 2004, Robertson and Richards 2003, Dorrian and Rose 2003, Hirsch and O'Hanlon 2003, Atkins et al. 1998).

Landscapes as materialized system include verbal, visual and physical aspects of human existence and the earth's surface, and create an arena for multi-dimensional and dynamic phenomena in the contemporary world (Cosgrove and Daniels 2004). The same time landscape interfaces various social categories like human/nature, time/space, past/future, expert/layperson and creates very broad field of multi-scale and multi-aspect discourse (Widgren 2004). Making landscape discourse more complex, some researchers (see Fægri 2004) argue, that cultural landscape can only be understood by its antithesis: untouched, unspoiled nature. But there is hardly any nature 'untouched' by humans, sometimes just by connotations and given meaning. It seems that technically everything on the planet has got some kind of human indication, no less than cultural significance. On the other hand, voices are being raised (see Cresswell 2004) suggesting that the term landscape should be abandoned, as being too common, too often used, as it is involved in too many explanations and interpretations. The variety of understandings and explanations of landscapes often brings misunderstanding, confusion and upheaval, which on one hand does not help to communicate and apprehend representatives of different 'schools' or approaches, but on the other hand can be very fertile and abundant source of interdisciplinary comprehensions.

Diversity of landscape interpretations

Complex and multi-aspect studies of landscape have recently made the term very popular and use in many disciplines, including both physical and human geography, architecture and urban studies, anthropology, social studies and ecology. The term 'landscape', however understood, has been frequently used not only in hundreds of scientific publications, but also in popular media and public discourse. Most of the participants of the landscape debate would agree that landscape comprises of visible features of an area of land, including physical elements such as landforms, living elements of flora and fauna, abstract elements such as lighting and weather conditions, as well as human artefacts and built environment. Landscape studies are generally based on recognition of the external world and comprise a complex and multi-aspect process, consisting of perception, but also comparison, classification, and valuation. This process often modifies and conditions our behaviour, and then our interpretation of landscape (Massey 2006).

Landscape usually refers somehow to the eye and a way of seeing, and is often understood as the outcome of a representational practice that stages its referent in relation to a viewing subject (Sennett 1990, Dorrian and Rose 2003). Landscape can be then recognized as total regional environment; a countryside; a land use; a topography or a landform; an ecosystem in which ecological relationships as realized as different types of landscape; as heritage or historical artefact; as a composite of physical components; an art form; a resource, and sometimes as energy and *genre de vie*. The types of contemporary landscape designations can be grouped into three main synthetic categories weighted according to the three basic attributes of landscape: form, use or significance (see Widgren 2004).

Landscape as form or artistic imagination

Landscape as form and visual aspects of a limited surface of earth is probably the most commonplace usage of the term, and comes directly from the *landschap* concept. The overall visual appearance of a stretch of countryside becomes a portion of land that the eye can comprehend in a single view. The visual landscape is often connected with its artistic imagination, and depicts scenery such as mountains, valleys, trees, rivers and forests. Sky is almost always included in the view, and weather usually is an element of the composition. The European artistic tradition of landscape imagination dates from the early 15th century, when landscape painting was established in Europe as an appreciation of natural beauty and part of spiritual activity. In Clark's (1949) analysis, there were four fundamental approaches to convert a material landscape to an idea: by the acceptance of descriptive symbols, by curiosity about the facts of nature, by the creation of fantasy to allay deep-rooted fears of nature and by the belief in a Golden Age of harmony and order. Although contemporary European landscape painters like Ton Schulten, Gabriella Benevolenza and Russell Frampton all play with bold colours and cubistic forms, they always include sky, earth and line of horizon in their imagined landscapes. At the same time, the Chinese tradition of purist landscape has been based on fine ink lines on silk or paper marking the outlines of hills and trees and expressing the lightness and simplicity of Buddhism (Andrews 2000).

There are two very important components of landscape as artistic imagination that are essential for further landscape interpretations in any other discipline: a theme that must be focused on a section of land and, no less important, there must always be a certain viewing point, or an imagined standing point of the painter or interpreter. The position of the artist structuralizes the scene and brings certain order and beauty to the painting. We must then remember that every 'objective' landscape imagination we see, we see only through the eyes and mind of the artist or decoder, and the standing position conditions to a great extent our points of view.

The other group of landscape interpretations comes from the art of landscaping and mostly refers to the process and practice of designing and planning large scale garden establishments. Initially a skill of classical times, landscape architecture – as landscape gardening – was rediscovered in Renaissance. Since that time its scope has broadened to cover the planning and management of landscapes created by human activities, but mainly as a mélange of nature and culture in a form of deliberate garden and open areas. The object is to arrange and position the created landscape into both functionally and aesthetically satisfying parks and gardens (Andrews 2000). Horticulture or the art of landscaping gardens has dominated the meaning of the word 'landscaping', the contemporary understanding of which is usually a connection of external beauty and aesthetic arrangements of forms, with the utilitarian, healthy aspects of gardens and parks.

The third designation of 'figurative' landscape, based on its visuality, comes from architecture. In recent years, landscape has emerged as a model for urbanism, which supplants the basic architectural designs of buildings and spaces between them (Waldheim 2006). Looking for a more holistic and up to date term, many architects use 'landscape' to emphasize that they design much more than just a building or its surroundings, but also create the whole visual appearance of the place or space. A cityscape then becomes the urban equivalent of a natural landscape, with 'townscape' roughly synonymous with 'cityscape', though it of course implies the same difference in urban size, density, and even modernity, implicit in the difference between the words 'city' and 'town'. In urban design, 'landscape' refers to the configuration of built forms and interstitial space. 'Landscape architecture' is a broader term which embraces landscape management, engineering, assessment and planning and generally requires a license and is concerned with large scale projects implemented over a long period of time and a wide geographical area (see Shannon 2006, Wolff et al. 1998).

Landscape as function or territorial complex

In natural sciences, and especially physical geographies, landscape is often understood as a geographical structure: a territorial complex consisting of soils, vegetation, hydrology, climate, as well as human settlements and communication network. The initial meaning of the word *landschap* as region or area has been employed in landscape ecology. Landscape or *landskap*, traditionally based on natural regions, can be also an administrative unit in Nordic countries, like Sweden and Norway, where landscape or cultural landscape research is habitually based on historical and regional analysis of rural settlements (see Birks et al. 2004).

Urban landscape can be of important value to the local economy and society. Landscape can be seen as a feature and function of a place, as well as a development option. The well managed and consumed landscape can enhance the quality of life, increase residential, investment and tourism attractiveness. Oscar Wilde, quoted in an office tower advertisement, said that 'it is only shallow people who do not judge by appearance' (Dovey 1999). When understood as a form of a function, the landscape is, in Harvey's (1982, 233) words, 'a geographically ordered, complex, composite commodity' that is fixed in the space and thus circulate more freely. The fixed environment 'functions as a vast, humanly created resource system, comprising values embedded in the physical landscape, which can be utilised for production, exchange, and consumption'. Landscape as a product can be then valued, sold, exchanged, explored, but also protected and treasured. This particular and compound resource is often employed and used in tourism and the real estate market.

More recently, landscape ecological approaches rose from growing concerns about the state of the environment. Since the 1970s in many European research centres, especially in Netherlands, the new methodology of landscape research embraced natural, ecological, cultural and social issues, and has been facilitated by the availability of new technologies, including remote sensing and GIS (Shaw and

Oldfield 2007). Landscape ecology developed in Europe from historical planning in human-dominated landscapes. In North America, concepts from general ecology theory were integrated. Landscape ecology, as a term introduced by a German geographer Carl Troll in 1939, is based upon heterogeneity in space and time, and frequently included human-caused landscape changes in theory and application of concepts. Landscape ecology is focused on spatial variation in landscapes, while landscape becomes merely a scale or equivalent of the micro-region. It includes the biophysical and societal causes and consequences of landscape heterogeneity and typically deals with problems in an applied and holistic context (Forman 1995). Landscape ecology looks at how spatial structure affects organism abundance at the landscape level, as well as the behaviour and functioning of the landscape as a whole. This includes the study of the pattern, or the internal order of a landscape, on process, or the continuous operation of functions of organisms. This interdisciplinary approach to the study of the environment at a landscape scale is essentially concerned with biological territorial capacity, landscape unit, ecosphere scale, landscape ecological principles, variegation model, and standard habitat (Ingegnoli 2004). Landscape ecology theory includes the landscape stability principle, which emphasizes the importance of landscape structural heterogeneity in developing resistance to changes, recovery from disturbances, and promoting total system stability (Forman 1995). This principle is a major contribution to general ecological theories which highlight the importance of relationships among the various components of the landscape.

Landscape as a territorial unit can be also studied as land use with all the economic applications and consequences. Analysis of land use changes has included a strongly geographical approach within landscape ecology. This has lead to acceptance of the idea of multifunctional properties of landscapes. A territorial complex and its visual attractiveness can be seen as a resource and market valued commodity. The collective landscape is a public good which should be protected and enhanced by legislation and public administration. Adopted in Florence, Italy, on 20 October 2000, the European Landscape Convention aims to promote the protection, management and planning of European landscapes and to organize European co-operation on landscape issues. It is the first international treaty to be exclusively concerned with the protection, management and enhancement of European landscape. The Convention emphasizes the significance of landscape:

> The landscape has an important public interest role in the cultural, ecological, environmental and social fields, and constitutes a resource favourable to economic activity and whose protection, management and planning can contribute to job creation. (The European Landscape Convention, 2000, Preamble)

Protecting existing public natural assets, societies are responsible for the creation of new public goods, and this can be done by positive landscape planning. Poland is one of the European countries where active 'landscape policy' and legislation has been implemented since the 1970s. The Polish school of landscape ecology, based on German, Dutch and Russian traditions, can be seen as one of the best examples of applied landscape ecology. The first Landscape Park in Poland was established in 1976 to restrict investment pressure for attractive, usually suburban areas or landscapes, and presently there are three forms of landscape protection: Landscape Parks, Areas of Protected Landscapes and Nature – Landscape Complexes (Wołoszyn 2006, Rychling and Solon 1998).

Legal solutions, connected with economic values and market regulations, impose strict and precise definition of landscape. Corresponding to the European Landscape Convention, landscape for European lawyers, ecologists, politicians and developers simply means 'an area, as perceived by people, whose character is the result of the action and interaction of natural and/or human factors' (The European Landscape Convention 2000, Chapter 1, Article 1).

Landscape as meaning or system of communication

The built landscape consists of network of the social relations that make and use it. While landscape signifies the look of the land, it also signifies a specific way of looking at the land. The landscape has developed as

> a way of seeing, a composition and structuring of the world so that it may be appropriated by a detached, individual spectator to whom an illusion of order and control is offered through the composition of space according to the certainties of geometry. (Cosgrove 1984, 55)

In anthropological terms, landscape refers to the material manifestation of the relations between humans and their environments. It is a product of the dialectic of physical environments and culture. Landscape is considered to be the social, economic and spatial background of human activities, which consist of a network of institutions, rules and laws, social order and representations. It is also a message, formulated by the ruling class to the others. Landscape then is a specific, spatial and grand scale signifying system, connecting both the *signifier* and the *signified* (de Saussure 1974). Landscape is an outcome and medium of social interactions and an input to the specific relations of production and reproduction. In recent years, landscape has been deployed as a framing convention or as a meaning imputed by people to their cultural and physical surroundings; it has became a feature between place and space, as defined by Hirsch (2003), and has been recently employed in various anthropological, cultural and social studies.

Landscape is not only divided into named land tracks and settlements sites; it is also and maybe above all, seen as structured by history. Moreover, it entails a relationship between the 'foreground actuality' and 'background actuality' of social life. Urban or rural scenery, as product of human values, meanings and symbols, reflects powers, needs, aspirations, as well as glorious and tragic history, written into the palimpsest of symbols and signs (Winchester, Kong and Dunn 2003). Urban landscape projects and communicates the view of the dominant element of society to the remainder, through the symbols written into the setting. Landscapes, then, reveal, represent and symbolize the relationship of power and control out of which they have emerged (Zukin 1993). Landscape is socially produced insofar is centred in relation to human agency. Cultural landscape seems to answer the demand for linkages between humans and their environments. As Tilly (1994, 10) writes:

[C]entred and meaningful space involves specific set of linkages between physical of the non-humanly created world, somatic states of the body, the mental space of cognition and representation and the space of movement, encounter and interaction between persons and the human and non-human environment.

Those linkages can be seen, analyzed and interpreted as integral elements of cultural landscape: landscape that is a transmitter of information, as well as value and meaning on its own. Landscape can carry many types of communications: an identity, a political statement and a scene for social rituals. Good example of landscape as message can be the Memorial of the Murdered Jews of Europe in Berlin. It can symbolize both the Jewish cemetery or/and the empty, concrete city. But cognition and representation is always based in the eye and mind of the observer, and for many visitors, the message intended by designers remains unclear, while the Monument becomes only an interesting and bizarre landscape feature and another tourist attraction.

Since the beginning of civilizations, architecture and urban landscape have been considered as clear and permanent transmitters of cultural codes. Urbanized space is a result of various locally and globally conditioned factors and forces. One of the oldest, Roman descriptions of the city defines it as a congregation of buildings, and of people who were able to create a number of public spaces and features symbolizing the common values and styles of the inhabitants of the city (Bielecki 1996). The form and significance of every piece of architecture or design reflects the internal, spiritual space of its creators, designers or investors, while the intentions and thoughts are more or less clearly mirrored in aesthetic form of spatial structure of the city. Recently, landscape as a communication system has been, up to a certain point, adopted by many geographers, sociologists and architects.

Since, as it was pointed out by Williams (1982), culture is embedded in other systems as constitutive component, landscape grows to be one of the most stable and significant cultural compositions. Most of the constituents of cultural landscape play not only strictly utilitarian, but also symbolic roles as well. One of the most common components of cultural landscape – dwelling – primarily satisfies the basic need for shelter, but beyond this, within the context of a particular society, dwellings signify a particular kinship or family system and further signify internal social differentiation (Williams 1982). The advantage of analyzing culture as a signifying system is that it emphasizes both the systematic quality of culture, as a structured system of signs, and its processual quality as something which is very temporal, contested and reaffirmed (Duncan 1990). In a very broad sense, landscape can be understood as the way in which people comprehend and engage with the material world around them. People's being-in-the-world is always historically and spatially contingent, so landscapes are always in process, potentially conflicted, untidy and uneasy (Bender 2001, Darby 2000). Cultural landscape then can be a synthesis of history and space, of memories and places, connotations and material forms, always changing and re-contexted.

The three above-mentioned mainstreams of landscape understandings are being supplemented by numerous detailed and particular definitions and interpretations. The multifaceted and complex notion of landscape, together with hundreds of definitions, does not make the landscape discourse easier to embrace or comprehend, but makes landscape almost a universal term rich of layers, meanings and explanations. Despite the various disciplines engaged in the landscape debate, geography seems to have the most comprehensive and developed practice in landscape studies.

Traditions of landscape discourse in geography

The term 'landscape' signifies the specific arrangement or the pattern of 'things on land'. It refers to the look or the style of the land, its role and the social or cultural significance of its order or make-up (Meining 1979). Landscape is not only a sphere of everyday life, of real values and practices, but it is also an important, and probably the most important, geographical source of meanings (Massey 2006). The history of geographical studies of landscape is as old as geography itself. Plutarch's *geō-graphía*, 'describing the earth', was actually not much different than contemporary descriptive approaches to landscape interpretations. People have been for centuries 'describing the earth', analyzing various manifestations of nature and cultures. The geographical tradition of landscape studies comes from the very core of the science, aiming and attempting to describe, understand, explore and examine the earth's surface, its visible and hidden components. A substantial part of the geographical analyses has practically focused on, in a very broad sense, landscapes investigation, in physical, economic, social and cultural spheres.

The geographical theory of landscape studies has been developing since the end of the 19th century, but even before this, the term 'landscape' had been constantly evolving and has been used in many connotations and contexts. Transformations and alterations of the understanding of landscape in geography follow general trends in development phases and shifting paradigms of the science. One of the first and very common definitions of 'landscape' in geography comes from the early 20th century German geographer Albrecht Penck, who defines geographical landscape as a limited area displaying common features (Shaw and Oldfield 2007). Geographers have long understood landscape to be just a built morphology – the shape and the structure of a place. This very plain definition has been adopted, modified and transformed by many schools and researchers, while shape and structure had been replenished by meanings and connotations. There are a few main traditions in geographical landscape explanations, connected with a range of philosophical approaches. The so-called main historical 'schools' of geographical landscape studies, including German, French, American and, least known, Russian/Soviet, influenced or sometimes dominated the landscapes discourse in geography during last 100 years, and to some extent still inspire and stimulate landscape researchers, not only within the field of geography.

German school of landscape studies

The bases of contemporary geographical studies of landscape had been laid by the German school of cultural geography. At the start of the 20th century, many geographers argue that landscape, as a part of the general biophysical environment, determined the cultures which existed within it. Simultaneously with the birth of the German nation-state, the German school emerged in the 19th century under the impulsion of Immanuel Kant's geography. For most of the German geographers of the time, the idea of nation was intertwined with that of culture. Both are considered souls and expression of an entire people. Culture feeds the dream of national unity and glory, as well as search for identity. The main topic of interest of cultural geographers is the relation of collective beings with their natural environment, or landscape. The main research topics included the meaning of differences on the surface of earth, relations between cultural and environmental differentiations and national path dependencies (Bonnemaison 2005).

The foundation of German school in human geography was laid by a number of philosophers, travellers and naturalists, interested in spatial differentiations. Gottfried Herder (1744–1803), the originator of the modern concept of 'nation', emphasized the role played by cultural communities in the construction of landscape and spaces. Herder sees the aim of geography as recreating the spatial and natural framework of people's history. The uniqueness of societies comes from their fusion with the earth and are scattered with geosymbols. People express their uniqueness by using the environment optimally, by creating a landscape specific to its culture and by letting the space/nature influence the culture. The naturalist philosopher Carl Ritter (1779–1859) extends Herder's intuitions and emphasizes the determining influence of the natural environment upon the origin and evolution of civilizations. He also articulates the essential dialectics of spatial scales by envisioning the municipality, then the region, the country, the nation, the cultural area and finally the world. Focusing on diversity of people and landscapes, the German school had been influence by social Darwinism, where societies are perceived as quasi-biological organisms that compete with each other (Bonnemaison 2005).

Early German practice of landscape studies had been influenced by environmental determinism, also known as geographical determinism. This approach is based on the view that the physical environment, rather than social conditions, determines human behaviour and cultures: humans are believed to be strictly defined by stimulus–response and cannot deviate. Similar ideas continued to be propounded until the modern era. Environmental determinism rose to prominence in the late 19th and early 20th century when it was taken up as a central theory by the discipline of geography, and to a lesser extent, anthropology. The fundamental argument of the environmental determinists was that aspects of physical geography, particularly of climate, influenced the psychological mind-set of individuals, which in turn defined the behaviour and culture of the society that those individuals formed (Winchester, Kong and Dunn 2003).

Probably the most famous representative of the German school of landscape studies was Friedrich Ratzel (1844–1904), who introduced the term 'anthropogeography' (*Anthropogeographie*), as a field of integration or merge of culture and nature, creating *Kulturlandschaft*, or cultural landscape. That idea gave rise to a romanticized view of landscape, closely connected with nationalism and Herder's concepts, but also created a background or reference point for many 20th century cultural geographical studies. Ratzel, the founder of German geopolitics, sees space as a place of power, a material incarnation of the Nation and State, and the true corporeal soul of the people. Using the idea of landscape as an expression of culture, geography becomes the science of landscapes or *Landschaftskunde*. The original meaning of German *Landschaft* is closely related to 'region', not quite in a sense the term 'landscape' is used in modern English. The definition of the word *Landschaft* refers to the symbiosis between landscape, as we understand it now, image and region; it means a defined geographical region in German – either a specific area or a type of area. *Landschaftskunde* falls in between contemporary 'landscape studies' and 'regional geography' (Bonnemaison 2005, Mitchell 2001).

One of the founders of landscape geography in Germany was geomorphologiest Siegfried Passarge (1867–1958), who argued that landscape should be seen as central device for the systematic organization of both physical and human data gathered in an area for scientific analysis. In his works, Passarge developed his ideas on landscape types and classification and attempted a major regionalization of the world, in which differences in natural vegetation become most visible (Shaw and Oldfield 2007). The fundamental concern of *Landschaftgeographie*, or geography of landscape, was with the landscape morphology, involving the examination of all that was visible on the earth's surface and the investigation of the characteristic associations of phenomena which existed in a specific region. The German school of landscape studies, together with initiatives and input in political and cultural geographies, established a base for all other trends and developments of modern landscape research. All the other schools or approaches had been constructed in opposition or as improvement of the old, early 20th century German landscape concepts.

French Vidalian geography

France, similarly to other nations, entered the modern era of geography shortly behind Germany. While the German school was filled with numerous scholars, the French owe much of their tradition to one man, Paul Vidal de la Blache (1845–1918). Although he lacked some of the spatial training of his German counterparts, Vidal would be given credit for helping to establish an entire generation of geographers. Vidal developed his own approach to geography, focusing on a regional method rather than a systematic one. His view of geography can be described as one steeped in chorology, unconcerned with the abstract debates of geography's scientific lineages and alignments (Hilkovitch and Fulkerson 1997). As Vidal puts it:

> ... that which geography, in exchange for the help it has received from other sciences, can bring to the common treasury, is the capacity not to break apart what nature has assembled, to understand the correspondence and correlation of things, whether in the setting of the whole surface of the earth, or in the regional setting where things are localized. (Martin and James 1993, 193)

While he recognized that geography was related to the sciences, both natural and social, neither was as important to geography's identity as its connection with human activity. He stressed the importance in viewing man in relation to his natural environment. As a regional geographer, he preferred to view man within a physical milieu, or *pays*, or within his cultural milieu, or *genre de vie*. Vidal applied the blend of culture and nature to the traditional French model of national territoriality with its regions or *pays*. Those ancient localities were considered to be geographical beings that drive from a combination of physical laws, bio-geological principle and human realities (Hilkovitch and Fulkerson 1997). The diversity of space illustrates the juxtaposition of regional structures, which is what geography investigates. Each region is based on a natural environment to which humans adapt themselves by using natural element as intermediary. This selection process gives a rise to a *genre de vie* or lifestyle. A *genre de vie* is what a group selects from a variety of natural elements in order to create a favourable life environment and build its culture. It includes the traditions, institutions, language, habits, foods, and so on. In other words, when ones studies geography regionally, one may identify a region or *pay*, with regard to its physical characteristics or with regard to its human characteristics (see Mitchell 2001, Bonnemaison 2005).

Vidal is also credited with providing a regional theory which highlights the man–land relationship with regard to personal and cultural development. Possibilism in cultural geography is the theory that the environment sets certain constraints or limitations, but culture is otherwise determined by people. The possibilist school of thought would believe that desert cultures would have similarities due to living in environments that have similar limitations, but they would also have important differences due to being different people. Possibilism is seen as Vidal's refutation to environmental determinism. Vidal acknowledges that environment plays a role in setting limitations and offering possibilities for personal and cultural development, but points out that humans can selectively respond to any factor in a number of ways. How people react and develop is a function of the choices they make in response to their environment. These choices are the manifestation of their culture, or *genre de vie* (Hilkovitch and Fulkerson 1997). Thus, possibilism is seen as an alternative to environmental determinism, explaining that how humans respond to their environment is in part a function of that location and partly a function of their own will.

Vidal's efforts came to be known as *la tradition vidalienne*, and soon after his death, the burgeoning field of geography was filled by many of students and followers throughout France, Europe and North America. *La tradition vidalienne* can be seen as a way of doing geography, marked by its emphasis on regional study and attention paid to the man–land relationships. The critics of the French school say that Vidalian region-place, based on *paysage*, seems to underestimate the social reality and creates a misconception regarding the determining role of politics. The concept additionally minimizes the role of towns and cities, since Vidalian regions were primarily rural (Bonnemaison 2005, Mitchell 2001). French geographers gave much thought to the concept of place, milieu, and lifestyles, emphasizing the fact that a place has naturalistic, historical and cultural dimensions. The Vidalian studies of *paysage*, the French equivalent to landscape, had been strongly influenced and limited by possibilism, but also opened new horizons for human geographical interpretations of regions and their visible phenomena.

Carl Sauer and the Berkeley School

In early 20th century America, many of the old European concepts, including schools in geographical thoughts, were considered less relevant. Landscape was, contrary to that of the Old Continent, not as culturally differentiated. Although the natural environment of the US undoubtedly varied from region to region, it did not allow for 'natural regions', as was the case in France or Germany. Pioneers had turned the past into *tabula rasa*, and pre-Columbian significant spaces had vanished. There remained an unlimited and monotonous space, an area showing no internal differentiation. The only distinction in space was brought by its use or function. The functional approach of economic landscape was further developed into the organization of monofunctional regions or belts, based on the optimal use of space (Winchester, Kong and Dunn 2003, Robertson and Richards 2003). Nowhere was it more visible than in the American Midwest, the homeland of Carl Ortwin Sauer (1889–1975).

The Berkeley school of cultural geography, established by Sauer, a professor at the University of California at Berkeley, flourished as a reaction to the materialistic and functionalist approach to landscape. In Sauer's eyes, productivistic determinism is much more dangerous than the pseudo-determinism of the natural environment. The best form of anti-determinism is culture, for culture is the very essence of unpredictability. Sauer explores geography as imprints of *genre de vie* onto landscapes, while culture is understood in its widest sense as the entirety of human experience, including spiritual, intellectual and material experiences. From the beginning of humanity, men and women have constructed their environment not only through a productivistic perspective, but also on basis of their values and representations. Cultural geography reconstructs the evolution of landscapes so as to elucidate their origins. To analyze a culture, Sauer defined visible elements of culture, which are linked with the material aspects of civilization and can be read in the landscape, where they play the role of markers that define levels of culture areas. These visible elements are connected with specific values, beliefs and rituals. As cultural markers, visible elements allow for the study of cultural evolution in space and time, which led Carl Sauer to investigate the diffusion processes. Languages, belief systems, customs and religions weave links between humans; these links leave a material trace. Landscape then becomes the matrix of identity as well as its imprints. Cultural trains are often referred into very visible signs, such as *toponymies* and *geosymbols*, that mark out a territory with an array of significant monuments, statues and so on. Geosymbols and cultural trails merge into cultural ensembles (Mitchell 2001, Sauer 1925).

Cultural landscape was generally understood and defined as the human-modified environment, including fields, houses, churches, highways, planted forests and mines, as well as weeds and pollution. Cultural landscape was defined by Sauer (1925) as a geographic area, comprising both cultural and natural resources and wildlife. The vicinity can be cultural only if associated with an historic event, activity, and person or exhibiting other cultural or aesthetic values. He further writes:

> The cultural landscape is fashioned from a natural landscape by a cultural group. Culture is the agent, the natural are the medium, the cultural landscape is the result. Under the

influence of a given culture, itself changing through time, the landscape undergoes development, passing through phases and probably reaching ultimately the end of its cycle of development. With the introduction of a different, alien culture, a rejuvenation of the cultural landscape sets in, or a new landscape is superimposed on remnants of the old one. (Sauer 1925, 46)

Sauer was explicitly concerned with countering an environmental determinism which had dominated the American geography of the previous generation, within which human agency was given inadequate power in the shaping of the visible landscape. Sauer stressed the role of culture as a force in shaping the visible features of the Earth's surface in delimited areas. Within his definition, the physical environment retains a central significance, as the medium with and through which human cultures act. As a critic of environmental determinism, he proposed instead an approach called 'landscape morphology' or 'cultural history'. This method involved the inductive gathering of facts about the human impact on the landscape over time. He drew heavily on the *superorganic theory of culture*, which saw culture as a causal agent sweeping individuals along with it. Sauer was particularly interested in the influence of cultures on material aspects of landscapes. He also delineated geography of milieux, with an analytical focus on the environmental characteristics of human societies (Bonnemaison 2005, Mitchell 2001).

Sauer's work aimed to map the distribution and dispersion of cultures across space. He was also interested how culture spread through regions, how some cultural trails displaced others and the relationship between the natural landscape and culture. There was a certain level of confidence that cultures were distinct, static, and therefore predictable. American cultural geography in the Berkeley tradition produced exhaustive surveys and atlases of the forms and material of houses, barns and fences (Winchester, Kong and Dunn 2003). To some extent, Sauer and his followers replaced environmental determinism with a specific cultural determinism. Not only did culture determine landscape, but culture also determined the individual. This determinism has been referred to as a superorganic conception of culture. This approach adopts the view that culture is an entity at a higher level than the individual that is governed by logic of its own and that it actively enables and constrains human behaviour. The superorganic concept sets culture apart from the environment: culture as a totality was imprinted on people and through them the landscape, ignoring the internal heterogeneity within a cultural group (Winchester, Kong and Dunn 2003). The Berkeley school of cultural geography was mainly focused on artefacts, objects and any physical aspects of culture, as well as on a limited definition of cultural group, practically restricted to ethnic or folk groups. Despite all of the limitations of the Berkeley school, the Sauerian methodology remained the main approach in cultural landscape studies until the late 20th century.

Russian landscape science tradition

Since the early 20th century, landscape science has played an important role in Russian, and later Soviet physical geography, occasionally aspiring to cross the physical–human divide. Together with 'general physical geography', landscape science constitutes a Russian 'national school' of geography. The Russian approach has emphasized landscape's bio-physical characteristics and its potential for utilization or transformation by humanity (Shaw and Oldfield 2007).

The Russian school of landscape science was initiated by soil scientist Vasilii V. Dokuchaev (1846–1903), whose applicative works on methods for combating drought and soil erosion in the steppe zone and for improving and regulating agriculture, forestry and water use provided a foundation for a distinctive Russian tradition in the environmental sciences. The human dimension of landscape was incorporated by Vasilii P. Semyonov-Tyan-Shansky (1870–1942), who insisted that the study of landscape should concern not only biophysical characteristics but also its visual, aesthetic, and cultural dimensions. Geography, according to Semenov, was close to art: studying landscape as *paysage* and fostering sensitivity to its colours, sounds and smells. This in a sense very post-modern, humanist and holistic attitude towards landscapes studies was not welcomed in totalitarian, Stalinist Soviet Union, where landscape studies were to distance society from environment and envisioned the domination, utilization, and manipulation of natural features (Shaw and Oldfield 2007).

In the mid 20th century, in the Soviet Union, distinct dimensions of landscape studies had been developed with geography. Landscape science or in Russian *landshaftovedenie* (literarily: *landscapology*) came from the German tradition of *Landschaftskunde* and is related to contemporary regional physical geography. The immense territory of Soviet Union and its enormous physical differentiation, as well as ideological limitation of social and cultural research, focused geographical and landscape studies on 'natural-territorial complexes', also called as landscapes. This 'territorial or aquatorial natural complex' was defined as

> a part of territory or water reservoir limited by conventional vertical borders according to relative homogeneity and horizontal borders based on obsolescence of the factor used to defined that complex. (Armand 1975, 18)

The concept of landscape as natural complex was initiated by zoologist Lev S. Berg (1876–1950) by his research on natural zones, and later intensely investigated by numerous Soviet geographers, including Stanislav V. Kalesnik, David L. Armand and many others. For Berg landscapes are law-governed, repetitive groupings, not only of forms of relief, but also of other objects and phenomena on the earth's surface. A geographical landscape is that combination in which the peculiarities of relief, climate, water, soil, vegetation and fauna, and to a certain degree human activity, blend into a single harmonious whole, typically repeated over the extent of the given zone of the earth (Shaw and Oldfield 2007). Detailed typologies and classifications, as well as a mathematical model, had been developed to analyze and categorize a variety of complex, predominately natural landscapes. *Landscape zone* was widely considered as the main natural system operating on earth, which included atmosphere, hydrosphere, lithosphere and biosphere. Sometimes, landscape was replaced with the term *geocomplex*, carrying similar meaning.

To research landscapes transformed by humans and their civilizations, a new subdiscipline, *constructive landscapology*, had been developed. This applicative approach was focused on anthropogenic landscapes, with special reference to 'cultural landscapes'. The category, according to David L. Armand (1975, 278) includes 'arable lands – fields, orchards, meadows, pastures etc. producing crops over the regional average (what can be cultural in an unkempt farm?), cultivated forests, parks, settlements'. *Constructive geography*, as with any Soviet science of the time, was infused by political messages and goals, and then ideologically engaged. The aim of *constructive landscapology* was to care for and multiply natural resources and fight for the betterment of social existence (Gierasimov 1966). Cultural artefacts were limited to economic and utilitarian aspects of the landscape, in a way similar to the American functionalist approach. It must be added that economic or cultural aspects of landscape were on the margins of the officially accepted research fields, dominated by more practical naturalist sciences.

Cultural landscape research in new cultural geography

The most recent approach in geographical landscape studies is based on post-modern understandings of cultures and places, and results from the so-called 'cultural turn' in social studies. In recent years 'culture' has become one of the most widespread terms, appearing in many different contexts. Popularity of the idiom comes from its universality and almost unlimited capacity. Contemporarily, the meaning of culture spans from pop, high and sub-cultures, ministries of culture, being cultural, cultural anthropology, agriculture, through cults and culturist. Culture has become one of the fashionable keywords which seems to open and fulfil almost any subject and discourse. One of the main reasons of recent attractiveness of 'culture' is its holistic character and ability to synthesize various features of the research objects. Culture can, and sometimes does, embrace all human activity and experience (Jencks 2004). The broad definition of 'culture' meets post-modern demand for an elastic meta-idiom, used more and more often by journalists, writers and researchers. Like 'landscape', 'culture' is a notoriously elastic concept and often defined as a signifying system through which 'a social order is communicated, reproduced, experienced and explored' (Williams 1982, 13). It involves the conscious and unconscious processes through which people live in – and make – places and landscapes, by giving meaning to their lives and communicating that meaning to themselves, each other and the world beyond (Cosgrove 1993, Graham 1998). Cultural studies has as its initial empirical focus the ordinary, the banal, and the everyday, hardly considered relevant research topics before. These are used as entry points to discussions of social relations, exposing relations of domination and cultural oppression. Cultural geography positions human beings at the centre of geographical knowledge – human beings with their beliefs, their passions, and their life experiences. Today, since 'everything is cultural', there is a fashion about culture, sometimes turned into a kind of fetish, when the adjective 'cultural' becomes something fashionable and posh.

Cultural geographers of the 20th century had, ironically, little interest in culture, and turned their attention almost exclusively to the artefacts (Duncan 1990). While traditionally landscapes have been recognized as reflections of the culture within which they were built or as a kind of artificial spoor yielding clues to events of the past, only rarely were they recognized as constituent elements in socio-political processes of cultural production and change. Transformation in social sciences and especially in culture studies created new possibilities for cultural geographical interpretations of landscape.

Cultural turn

The cultural turn refers to the methodological and thematic shift in various disciplines of culture research, including cultural and political studies, sociology of culture, anthropology, philosophy, economy and geography. It describes a shift in emphasis towards meaning through culture rather than politics or economics. This shift of emphasis occurred over a prolonged period from the late 1960s, but particularly in the late 1970s and 1980s. With the move towards meaning, the importance of high arts and mass culture in cultural studies has declined. If before culture had been about things, like a piece of art, or a TV series, it has now become more about processes and practices of meanings. The cultural turn was a significant part of the post-modern turn or post-structuralist turn and created great ferment and stimulation, especially in English-speaking academia in the study of culture, and the theory and practice of interpretation associated with it. What the cultural turn has meant for anthropology is a strong reinforcement of its already strong valorization of hermeneutic depth in ethnography, and consequently its concerns with the interpretation of symbols, meanings and representations, at the cost of its treatment of social relations, social structures and systematic differences in economy. The cultural turn has helped cultural studies to gain more respect as an academic discipline. With the shift away from high arts the discipline has increased its perceived importance and influence on other disciplines (see Best and Kellner 1991, Johnson et al. 2004).

During last 20 years, contemporary cultural geography has produced, mainly in Anglo-Saxon countries, new ways of thinking about culture and geography (Ryan 2000). Older theories in many ways stressed time, while recently culture became much more spatial then ever before (Mitchell 2001). New cultural theory, as it has been developed in anthropology, sociology, philosophy and many allied disciplines, stressed space. Understanding of culture is constituted through space and as a space. What the culture turn has meant for geography is a strong intervention of interpretative theories, methods and ideas in a field heavily influenced by tasks of mapping, describing societies spatially, and by economic thinking (Marcus 2000). In geography cultural turn was mainly accompanied by or associated with post-modernist approaches. The differences between geography before and after cultural turn identified with post-modernity were well described by Mitchell (2001, 58):

> Postmodernism stresses heterogeneity, whereas modernism is accused of seeking homogeneity; postmodernism looks to multiply, competing discourses, whereas modernism seems to always want a single metanarrative capable to explaining everything; postmodernism is infatuated with the intermediacy of language, knowledge, and social practise and the elusive search for the meaning, while modernism stands mired in the overweening and impossible desire for self-contained explanation, for rational action in the face of perfectly knowable processes and actions.

Postmodernism is characterized by a strong emphasis on space; what adds to the post-modern and cultural turn of the 1980s is the new characteristic of a spatial turn. The language of space and place has become most popular and used in many different contexts and discourses. Spatial metaphors have become indispensable for understanding the constitutions of culture, as realm, medium, level and zone. The confirmation of the importance of space and, indirectly, geography, came from the eminent social philosopher Michel Foucault (1986, 22), who declare that 'the present epoch will perhaps be above all the epoch of space. We are in the epoch of simultaneity: we are in the epoch of juxtaposition, the epoch of the side-by-side, of the dispersed' (see Crang 2002).

According to Philo (2000), human geographers have gradually overcome their fear of the immaterial, of things without obvious material expression in the world, and have thereby opened up the possibilities for the cultural turn. Traditional human geography generally took the endpoint of its enquiries to lie in patient accounts of obvious, tangible, countable and mappable phenomena present to the senses of the geographical researcher. The process of *dematerializing* human geography is represented by

> preoccupation with immaterial cultural processes, with the constitution of intersubjective meaning systems, with the play of identity politics through the (…) spaces of texts, sighs, symbols, psyches, desires, fears and imaginings. (Philo 2000, 33)

Despite Philo's (2000) suggestions, this *dematerialization* of geography still worries and sometimes terrifies some of the experienced traditional researchers, who could not find 'science' and 'matter' in new, hardly countable and generally immaterial geography. The methodological turn is social sciences, either called 'post-modern', 'spatial', 'cultural', or as identified by Foucault, 'geometric turn' in the history of power,[1] gives research directions towards cultures and spaces, and had stimulated the development of the new cultural geography (see Crang 2004, Meining 1979, Mitchell 2001, Czepczyński 2007, Johnson et al. 2004, Best and Kellner 1991).

New cultural geography

In the 1980s cultural and post-modern turns had been gradually adopted in many geographical research centres, particularly in the US and UK. A clear distinction had been drawn within cultural geography between the traditionalist approach of the Berkeley school and what is now called the 'new cultural geography'.[2] The most

---

1 Foucault's 'geometric turn' might sound a bit peculiar and in relation to the contemporary geographical studies and can be actually called 'geographical turn'. It seems that Foucault had some reserve in using the term 'geography', as a field predominated by modernist methodologies in the 1970s and early 1980s.

2 The renewed discipline, now with 20 years of practice, has recently been losing its quotation marks, and from 'new' cultural geography is now more often called new cultural geography.

famous exposition of this dichotomy was by the British cultural geographers Denis Cosgrove and Peter Jackson (1987, 95):

> if we were to define this 'new' cultural geography it would be contemporary as well as historical (but always contextual and theoretically informed); social as well as spatial (but not confined exclusively to narrowly-defined landscape issues); urban as well as rural; and interested in the contingent nature of culture, in dominant ideologies and in form of resistance to them.

The simplest definition of new cultural geography is closely related to the new, broader and holistic definition of culture and is associated with cultural turn. New cultural geography can then be defined as the study of geographical aspects of human culture as processes and practices of meanings; all possible aspects of different spatial and thematic levels. Based on new and more vibrant and interactive understanding of culture grew humanistic geography or the geography of representations, as a newer aspect of cultural geography, where cultural space is wrought on the basis of representation. On this context cultural space is a space of belief in common values structures by ideas or ideologies (see Mitchell 2001, Marcus 2000, Crang 2004).

Cosgrove and Jackson's (1987) concept of new cultural geography was raised from criticism of the Berkeley school based on its narrow focus on physical artefacts, as well as a unitary view of culture rather than a constantly negotiated and constituted plurality of cultures. They also argued for a more complex concept of landscape, recognizing it as a cultural construction, a 'particular way of composing, structuring and giving meaning to an external world whose history has to be understood in relation to the material appropriation of land' (Cosgrove and Jackson 1987, 96). Landscape has been seen as a social and physical construction, where symbolic and represented landscapes produced and sustained social meanings, visualized in physical forms. In the pursuit of such analyses of landscapes, they highlighted the usefulness of two metaphors: landscape as text and landscape as theatre (see Cosgrove 1998, Cosgrove and Daniels 2004, Kong 2007). Cosgrove and Jackson's (1987) vision for contemporary cultural geography is multi-level and holistic. It would, moreover, assert the centrality of culture in human affairs. Culture is not a residual category, the surface variation left unaccounted for by more powerful economic analyses; it is the very medium through which social change is experienced, contested and constituted (Kong 2007).

New concepts and methodologies have developed renewed emphasis on 'doing' of cultural geography (Shurmer-Smith and Hannam 1994). From heritage to race to religion, issues such as multiplicity of meanings, cultural politics, ideological landscapes and constructions of identities have been explored. The new explorations of meanings have been conveyed in representations of places, landscapes and nature, especially in mass media, popular culture and advertising (Shurmer-Smith 2002). Since 'cultures are politically contested' (Cosgrove and Jackson 1987, 99), so are cultural landscapes. Landscape becomes a medium of competition and contestation between groups, evident in the requisition and conversion of significations from the dominant culture by subordinate groups as forms of resistance. In this view, the production of social knowledge can be recognized as uneven, with 'texts' as well as 'blanks' (Kong 2007). The call for more interpretative rather than morphological analyses draws attention to the importance of images in landscape analysis and therefore Panofsky's (1972) notion of iconography. As with textual metaphors, images hide multiple layers of meanings and are not 'transparent windows' to the real world (Cosgrove and Jackson 1987, Kong 2007).

The transformation of cultural geographies, as initiated or recapitulated by Cosgrove and Jackson, found many followers and supporters. The vast majority of contemporary cultural geographical studies apply basic concepts and notions of new cultural geography. Contemporary societies of the post-modern world, where signs and symbols are inverted and recycled in different contexts, do not reassure the stability of meanings but instead freedom of intertextuality and interpretation. This, as many argue, results in surface rather than depth meanings, which calls to question the usefulness of new methods in cultural landscape research (Kong 2007).

Landscape studies in new cultural geography

Landscape is one of the central elements of a cultural system, for as an ordered assemblage of objects, it also acts as a specific representative system. In order to understand this structured and structuring quality of landscape, we must first inquire into what is or can be signified by landscape. Duncan (1990) calls this process signification of landscape. Later, we must examine the manner in which this signification takes place, or rhetoric of landscape. The first is an examination of local people's accounts of the nature of the landscape, what importance they attach to the landscape and how their readings of landscape contribute to politics of interpretation that either naturalizes the social relations in a society or transforms them. It involves the researcher's interpretation of what a landscape signifies to those who produce, reproduce or transform it. The hermeneutic problematic acknowledges the historical, cultural and intellectual frames of reference which the academic brings to bear on his or her interpretations and the role this must necessarily play in historical investigation. It also takes seriously commonsense beliefs, values and explanations. As Anthony Giddens (1974, 316) has said, these are not 'adjuncts to human action, they are integral to it'. He goes on: 'lay beliefs are not descriptions of the social world, but they are very basis of the constitution of that world, as the organised product of human acts'. All those common representations of culture are important significations, which typify hidden beliefs and thoughts.

It is hardly possible to present a holistic, comprehensive contemporary definition of cultural landscape, since both terms, culture and landscape, carry so many meanings, explanations and understandings, and the mélange of cultures and landscapes creates dozens of possible, used and interesting definitions. Cultural landscape has become the main cultural product of our time, as well as one of the crucial issues of post-structural geographies. Post-structural studies of cultural landscape are focused on two aspects of the phenomenon: landscapes as representation of cultures and landscape as an integral component of a culture. The interrelated arrangement of symbols and structures is based on interpretations. The signs or text may be transcribed on various levels, like form or architecture, use or function, meaning and representation, and on various bases, including aesthetic, ethical, political, architectural, historic, economic, financial, legal, infrastructural, cultural, social, semiotic, as well as environmental. Cosgrove (1993) describes cultural landscape as a complex social construction contested along multiple and overlapping axes of differentiation. In this context, all cultural landscapes are imaginary in the sense that they cannot exist for us beyond the socially constructed images which we form of them in our minds (see Shurmer-Smith and Hannam 1994, Graham 1998, Tuan 1990, Wallach 2005, Czepczyński 2006c).

The meaning of landscape is a compromise between the visible and the hidden, between reason and emotion, between morphology and function. The morphologies include historical glory; their monumental magnificence typically accentuated by usually ambitious imaginary (Turnbrige 1998). At the same time, cultural landscape is a picture of symbols rather than facts (Zukin 1993). It is also a product of the human values, meanings and symbols of the dominant culture within society. Landscapes as cultural products embody the culture of both the creators and the percipients. The setting can be represented and expressed by many modes of culture, including architecture, habits, literature, thoughts and meanings. Culture, in a very broad and post-modern sense, is central to understanding the landscape in that it frames and symbolizes economic, social and political processes. The compilation of cognitive objects and affective meanings forms the basis of new cultural geography, and are most crucial for the interpreting cultural landscape (Black 2003, Hall 2002, Zukin 1993, Czepczyński 2006b).

Personal attitudes and feelings towards cultural landscape are seen as a crucial component of post-modern culture. The inner landscape or the imprinted setting within us, oriented by a personal 'cognitive map', leads us and often determines our spatial behaviour. The liminality of the inner landscape finds an echo in the city's production of liminal space, incorporating areas that used to be tightly defined by social limits into market culture. The personally defined social landscape, based on private experiences, marks out the meaning and significance of the visible urban scenery. The inner or perceived landscape embodies our point of view. Our cognitive maps, aesthetic forms and ideologies replicate – while being reflected by – the shapes of the inner landscape (Zukin 1993).

Cultural landscape can be seen as being among the main composers as well as vivacious transmitters of culture. Forms and meanings are the bodily expressions of the ways of thinking, the experience, and the hierarchies of values and culture of each of the group as well as of each individual. This culturally conditioned *episteme* is always boldly visible through the forms of urban space and its organization. Cultural landscape is one of the main representing languages of modern society, which signify the spiritual dimension of the investors, architects and users. The surrounding is central to understanding the landscape in that it embodies economic, social and cultural processes. The aesthetic form is never neutral – power is written into landscape through the medium of design, usually used and overused by rulers to stress authority and legacy (Markus and Cameron 2002). Urban landscape is self-evidently a cultural symbol, however culture is understood. Landscape is a part of culture and expresses the needs, values and norms that shaped it in the past and maintain it in the present. Morphology of the city is thus a medium through which these attributes are transmitted, an artistic production expressing the past and present aesthetic values of the societies that deliberately created it (Ashworth 1998).

Cultural landscape studies are based on syntheses of events, features and processes that are visualized in urban or rural morphology, spatial structure and meanings, attached to buildings and places. Cultural landscape can be understood as both visual and symbolic picture of cultural values. The values are being reflected in forms, but also in social behaviour and individual activities, undertaken with certain spatial frames. Landscape as value, language and meanings constructs social, cultural and political reality. Cultural landscape can play many functions and can be interpreted on many various levels, including:

- *Landscape as personal identity* and environment, which makes the external framework of one's daily life. It is a function and a part of one's personality. Behaviourism, personal connotations and memories become most important within this function of landscape. Signs, symbols and meanings coded into cultural landscapes allow humans to situate themselves in time and space and to identify with a given culture and society. Psychological stimulation of landscape can be an activity inspiration factor, in both positive and negative denotation.

- *Landscape as symbolic exchange* is being employed into political and social representation systems. Landscape as a form of social control is packed with perpetual social values. It is also the realization of dreams, ideas and projects. Landscape is furthermore a heritage that belongs to the social collective memory of a particular group. The history of a place, nation, region, city as well as individual stories and memories are interwoven into the heritage landscape. A special role is played by politics and political interpretations of history, memories and landscapes.

- *Landscape as market resource* is related to the economic and profitable aspects of scenery. Attractiveness and market values of certain, popularly recognized features of landscapes becomes a marketable and sellable good. Land use and land rent turn into core foci of landscape as a resource discourse, while imagery of place, closely related to cultural landscape, is an important part of place marketing, used for tourism, investments and residential developments (see Bonnemaison 2005, Robertson and Richards 2003, Mitchell 2001, Czepczyński 2006b).

Landscape, particularly cultural landscape, plays a very important and often undervalued role in the estimation and valuation of human quality of life. The majority of external factors that influence our emotions come from the surrounding landscape. The emotional value of cultural landscape implies our everyday patterns and feelings. The appeal and attractiveness of surrounding urban landscape can positively or negatively influence our daily existence. Visible values of adjacent landscapes play an immense role in either increasing or decreasing the quality of everyday life. The idea of happiness is reflected in attitudes towards quality of life and the happy existence. The Epicurean positive balance of emotional experiences is often enriched by association with the external world, often seen as landscape. The opulent and assimilated cultural landscape, as a product of civilization, can supplement and enhance the residential milieu. 'Good scenery', where the pleasant areas predominate over the gloomy aspects of landscape, appears as an essential factor of living the 'good life' (see de Botton 2007).

Landscape is very seldom a stable or fixed-forever spatial structure. It is a cultural process that brings together the cultural meaning and the concrete actuality of everyday life. Cultural landscape can be recognized as a product of interrelations, as area under permanent construction and as sphere of the possibility of the existence of multiplicity, in which different trajectories coexist (Massey 2006). The dynamic and constructive nature of the surrounding environment implicates cultural landscape studies as multi-vocal and multi-factor reading of the concept. The post-modern 'world as an arena' metaphor seems to be the most appropriate attitude to understanding the urban landscape seen as a dynamic scene or theatre, an ongoing show with thriving *geo-stage-design*. One of the main goals of geographical research is to discover and interpret human spatial behaviours, patterns and relations. Space and landscape are important factors that condition our behaviours. Landscape is overtaken by its users, adjusted and constructed according to the socio-cultural standards used by local society. The relation between social structures and intentional spaces creates the frames for cultural landscape interpretations. Social creation of spaces and landscapes has both utilitarian and symbolic character. New constructed or re-constructed cities are based on clearly defined systems of existential, social, political and economic values (Rembowska 1998). Spaces, physiologies, and symbols of urban life are closely related to the ideological background of the dominant social, political and economic systems. Cities are planed, constructed or transformed according to specific designs and plans, inspired and controlled by local or global powers. In consequence, spatial, architectural, social and cultural structures, as well as features and details, are evident and obvious in urban landscapes and most of then can be de-coded and symbolically interpreted, using wide variety of methodologies.

Methodological pluralism

The basic sense of landscape, as a visual and horizontal surface with cultural meaning, is being interpreted in a range of methods, styles and modes, which express various emphases and other disciplinary backgrounds of the interpreter. The varieties of new cultural landscape interpretations are generally based on post-modern, or rather anti-modern, approaches and can employ an assortment of research methodologies (Johnson et al. 2004). Since the cultural landscape is a variable product of interrelations, sphere of possibilities and coexistence of different trajectories (Massey 2006), it might be rather difficult to understand and explain the complexity of landscape using just one methodology. Many researchers opt for a combination of supplementary attitudes and methods to synthesize the multifarious phenomenon of cultural landscape.

Given these perspectives on culture and landscape, there were concomitant and inevitable implications for methodological routes. Because of object fetishism, fascination with historical reconstruction and belief in the possibility of unmediated observation as a guarantee of objectivity, many human geographers have largely limited their research methods to observation and archival studies. As Cosgrove and

Jackson (1987) rightly pointed out, there was a need to adopt more interpretative rather than strictly morphological methods, and in line with the landscape as text metaphor, the most commonly favoured methods were drawn from linguistics and semiotics. In this view, the landscape was a text to be read as a social document or a spatial language, a parallel to Geertz's (1973) anthropological interpretation of cultural texts in which the idea was that there were multiple layers of meaning, including the one that social scientists added on through their interpretation. These layers of meaning were to be disclosed through a process of description. Related to this acknowledgement of multiple layers of meaning is the view from anthropology in particular, that ethnography too is a text, which 'calls into question the nature and history of the production of social knowledge itself' (Cosgrove and Jackson 1987, 97).

Post-structural approaches

Most of the post-modern social studies' methodologies are related to post-structuralism. In direct contrast to structuralism's claims of culturally independent meaning, post-structuralists typically view culture as inseparable from meaning. Structuralism studies the underlying structures inherent in cultural products, such as texts and landscapes, and utilizes analytical concepts from linguistics, psychology, anthropology and other fields to understand and interpret those structures. Post-structuralism is difficult to define or summarize because it rejects definitions that claim to have discovered absolute truths or facts about the world and very few people have willingly accepted the label post-structuralist; rather, they have been labelled as such by others (Best and Kellner 1991, Aitken and Valentine 2006).

Post-structuralists hold that the concept of *self* as a singular and coherent entity is a fictional construct. Instead, an individual comprises conflicting tensions and knowledge claims. Therefore, to properly study a text a reader must understand how the work is related to one's own personal concept of self. This self-perception plays a critical role in one's interpretation of meaning. The author's meaning is secondary to the meaning that the reader perceives. Post-structuralism rejects the idea of a text having a single purpose, a single meaning or one singular existence. Instead, every individual reader creates a new and individual purpose, meaning and existence for a given text. A post-structuralist critic must be able to utilize a variety of perspectives to create a multifaceted interpretation of a text, even if these interpretations conflict with one another. It is particularly important to analyze how the meanings of a text shift in relation to certain variables, usually involving the identity of the reader. Post-structuralism advocates deconstruction, a premise which claims that the meanings of texts and concepts constantly shift in relation to many variables. The only way to properly understand these meanings is to deconstruct the assumptions and knowledge systems which produce the illusion of singular meaning. Any mode of communication, including cultural landscape, has a certain vagueness which makes precise interpretation impossible. Interpretation, therefore, is equally in the hands of the reader and the author. To understand an object, it is necessary to study both the object itself, and the systems of knowledge which were coordinated to produce the object. In this way, post-structuralism positions itself as a study of how knowledge is produced. Post-structuralist studies often emphasize history to analyze descriptive concepts. By studying how cultural concepts have changed over time, post-structuralists seek to understand how those same concepts are understood by readers in the present (see Koch 2007, Harrison 2006).

Cultural landscapes can be understood as one of those cultural concepts, employing many perspectives to interpret and understand both the meanings of the authors, as well as contemporary interpreters. In view of the fact that cultural landscapes are generally speaking historical, descriptive analysis helps to decipher the production of knowledge coded with cultural landscape features.

Discursive studies

Discursive approaches are closely related to post-structuralism and can be defined as a social framework of intelligibility within which all practices are communicated, negotiated or challenged (Duncan 1990). According to the literature on discourses, ideologies are inscribed in them; ideologies inhere in the very language and the narrative structure of discourses. In semantics, discourses are linguistic units composed of several sentences – in other words, conversations, arguments or speeches. In the social sciences, a discourse is considered to be an institutionalized way of thinking, a social boundary defining what can be said about a specific topic. Discourses are seen to affect our views on all things; it is not possible to escape discourse. The chosen discourse delivers the vocabulary, expressions and perhaps also the style needed to communicate. Discourse is closely linked to different theories of power and state, at least as long as defining discourse is seen to mean defining reality itself (see Best and Kellner 1991).

This concept has been transferred into philosophy and is linked with the works of Michel Foucault (1926–1984), who was developing a battle-type of discourse which opposed the classic Marxist definition of ideology; Jürgen Habermas, who tried to find the transcendent rules upon which speakers could agree on a groundwork consensus, and many others. Since, according to Foucault (1975), knowledge and power are intrinsically related, power relations are immanent to discourses, whereas in the classic Marxist conception, the discourse is conceived as the ideological superstructure, but this does not impede the power relations being essentially located in the economic base, afterward reflected in the superstructure. Furthermore discourse is not anyone's property and thus has no essentialist meaning. The same discourse may change political sides quite often, being reappropriated and endlessly modified, as Foucault showed in his analysis of the historical and political discourse, there is a 'polymorphic tactics' of discourses. In other words, specific discourses are not tied to the subject; rather, the subject is a social construction of the discourse (Foucault and Gordon 1981, Foucault 1986, Best and Kellner 1991).

Foucault's (1975) discourse must both be understood as a singular discourse and as a more general discourse, meaning the boundaries given to any particular discourse. In this more general sense, discourse is not composed only of words, which would be to limit oneself to a dualist conception; discourse is also composed of *architectural dispositifs*, understood as heterogeneous assemblage, containing discourses, institutions, architectural buildings (*aménagements architecturaux*) or what we can call urban landscape, legal decisions, as well as statements and philosophical propositions. The *dispositif* can be both a verbalized and non-verbalized element, over and above the network that can be established between those elements (Foucault 1975, Foucault and Gordon 1981). These architectural *dispositifs* can be seen as spatial units or arrangements composed of several buildings, places and cultural connotations, and become equivalent to cultural landscapes. Discursive landscape provides certain modes of communication; a language in which power relations are conversed in polymorphic urban texts.

Subjectivity and situated knowledge

One of the main and basic research questions has being raised since Plato and Socrates' debates: what is real and how do we know what we infer about the real is true? Empirical evidence based upon observation and experimentation in the physical world is required for the verification of scientific judgment, and adherence to the rules of deduction and the process of inductive reasoning determines the validity and soundness of scientific argument and conclusions. Whether independent propositions exist as the objects of objectivism, or as the timeless truths concerning an object once it has become the intended object of a mental act, their reason for being would appear to be essential only to the process of discovery. The fundamental criticism then becomes one that is similar to the criticisms levied against, for example, historical objectivity. In any social research, we always select subjects and methodologies, and the selection is influenced or facilitated by arrangements of factors that are not always fully objectively verifiable. Taking an objective approach may not always be relevant, particularly in cases where it is impossible to be objective either because the relevant facts and viewpoints necessary are lacking, or because it is the subjective opinion or response that happens to be important. Thus it is possible to take an objective approach appropriately in situations which call for an expression of subjective thought or feeling. In this the problematic relation of truth to objectivity becomes evident. In reason, subjectivity refers to the property of perceptions, arguments, and language as being based in a subject point of view, and hence influenced in accordance with a particular bias. Subjectivity refers to the specific discerning interpretations of any aspect of knowledge, unique to the person experiencing them (Johnson et al. 2004).

By the end of 1980s the subjectivity of knowledge and research in the UK has materialized in a new approach called 'situated knowledge'. Initially this was mainly connected with feminist and gender studies, but since the 1990s situationism has appeared in other social sciences, including cultural geography. Its main position is based on the relation between object and result of research. In traditional scientific paradigms, one-way relation, in which the text, map or any research result, simply represented the object of study. Researchers operated as standardized conduits and their texts were considered simple matter of facts (Cook et al. 2005). Post-modernity highlighted the role of positionality and the situated nature of all knowledge, including academic. The simplistic object–text relation is replaced by a more complex correlation, giving a researcher and an audience an active role in text-making process. The object of study is researched and influenced by the scientist, who chooses the methods, approaches and positions according to his or her own experiences, beliefs, contexts and expectations. The final result of the 'positioned' research is a comprehensive interaction between the scientist, object, text and readers (Cook et al. 2005). The objective outsider's perspective is left behind in order to capture the native's or deeper and detailed point of view (Hirsch 2003, Valentine 2001). Interpretations of cultural landscapes always are 'situated', as any post-structuralist studies are. At the core of landscape as art interpretations lays the position of the observer; there is always specific viewing point or an imagined standing point of the interpreter. The situation and position of the landscape decoder structuralizes the scene each time: every objective-looking imagination is merely a projection of the artist or scientist.

Social construction theory

Social constructionism or social constructivism is a sociological theory of knowledge based on Hegel's ideas, and developed by Durkheim at the beginning of the 20th century. The concept has become more popular since the 1960s and has been widely applied in many social sciences (Berger and Luckmann 1966). The focus of social constructionism is to uncover the ways in which individuals and groups participate in the creation of their perceived reality. It involves looking at the ways social phenomena are created, institutionalized, and made into tradition by humans. Socially constructed reality is seen as an ongoing, dynamic process; reality is re-produced by people acting on their interpretations and their knowledge of it. Berger and Luckmann argued (1966) that all knowledge, including the most basic, taken-for-granted commonsense knowledge of everyday reality, is derived from and maintained by social interactions. When people interact, they do so with the understanding that their respective perceptions of reality are related, and as they act upon this understanding their common knowledge of reality becomes reinforced. Since this commonsense knowledge is negotiated by people, human typifications, significations and institutions come to be presented as part of an objective reality. It is in this sense that it can be said that reality is socially constructed (Cook et al. 2005).

A social construction or social construct can be considered as an institutionalized entity or an artefact in a social system invented or constructed by participants in a particular culture. According to social constructionists, constructs include such things as games, money, school grades, titles, nationalities, governments, universities, corporations and other institutions. Other social constructs might include language, class, race, gender roles, religion, science, childhood, sexuality, morality, profanity, memory and reality. Social constructions must be seen in an institutional context, as arising from the institutionalization of patterns of interaction and meaning in society leading to a construction of social institutions and institutionalized perspectives and understandings (see Hacking 1999). The notion of social construction has continued to feature prominently in the social sciences, especially in cultural and political studies. Cultures have been constructed and at the same time cultures construct us. Every element of human culture, like tradition, heritage, education, system of values and meanings, as well as cultural landscapes have been formed and facilitated by people and they are understandable and important only for the people of particular culture.

Landscape is always socially constructed, whether by intent or default. Truth, like art, is placed in the eye of the observer; the true landscape is also primarily in the eye of the participant or the viewer. The significance of objects, things and places is always culturally conditioned. Many scales and dimensions of personal and group identity reflect the contestation of societies along many various axes that include nationalism, regionalism, locality, class, gender, ethnicity, material well-being and political preferences (Graham 1998). The interpretation of cultural landscape and its features mirrors to a great extent the axis of differentiation, while nationalism and political preferences seem to be important in reinterpretations of the past.

Semiotics of landscape

Semiotics, semiotic studies, or semiology is the study of signs and symbols, both individually and grouped into sign systems. It includes the study of how meaning is constructed and understood. Semioticians classify signs or sign systems in relation to the way they are transmitted. To coin a word to refer to a thing, the community must agree on a simple, denotative meaning within their language. But that word can transmit that meaning only within the language's grammatical structures and codes. Codes also represent the values of the culture, and are able to add new shades of connotation to every aspect of life. Ferdinand de Saussure (1857–1913), considered the father of modern linguistics, proposed a dualistic notion of signs, relating the *signifier* as the form of the word or phrase uttered, and to the *signified* as the mental concept. It is important to note that, according to de Saussure, the sign is completely arbitrary; there is no necessary connection between the sign and its meaning and no word or sign is inherently meaningful. Rather a word is only a *signifier*, the representation of something, and it must be combined in the brain with the *signified*, or the thing itself, in order to form a meaning-imbued sign. De Saussure believed that dismantling signs was a real science, for in doing so we come to an empirical understanding of how humans synthesize physical stimuli into words and other abstract concepts (see Silverman 1998, Blonsky 1995, Jay 1998).

Lévinas (2003, 11–12) connects linguistics and semantics to the positionism and social constructions:

> ...language refers to the position of the listener and the speaker, that is, to the contingency of their story. To seize by inventory all the contexts of language and all possible positions of interlocutors is a senseless task. Every verbal signification lies at the confluence of countless semantic rivers. Experience, like language, no longer seems to be made of isolated elements lodged somehow in a Euclidean space... [Words] signify from the 'world' and from the position of one who is looking.

Semiotics differs from linguistics in that it generalizes the definition of a sign to encompass signs in any medium or sensory modality. Thus it broadens the range of sign systems and sign relations, and extends the definition of language in what amounts to its widest analogical or metaphorical sense (see Caravetta 1998). Landscape can be seen as a particular system of communication or language, where buildings and arrangements play similar role to the words and phrases. The statement coded in landscape might be compared to the verbal announcement, and often the textual information or slogans are transferred into the build and meaningful forms. The 'position of one who is looking' can be easily transferred into the cultural landscape situated studies. Semantics of landscape become one of the most important fields of cultural landscape studies, concentrated on decoding and interpreting meanings and symbols of the surrounding environment.

Interpretations of cultural landscapes

Space can be important for humans not only because its purely utilitarian values, but also because it reflects cultural and symbolic values. The meaning of space depends on cultural capital of society, as well as on many external conditions, including environmental, economic, politic, religious, which directly or indirectly influence the forms and significance of places and features (Norberg-Schulz 1999). Since the beginning of civilization power or significance has been mediated in spatial forms. Driven by social forces, the demand for instant fame and economic growth, the expensive landmark has recently challenged the previous tradition of architectural feature. In the past important public landscape and urban arrangements expressed shared meanings and conveyed them through well-known conventions. Some old towns retain these relationships of powers and meaning today: the tallest building may still be the parish church; the less prominent might be the school or public library; and minor civic buildings adopt a modest demeanour. There is a hierarchy of public worth, not perfectly agreed and finely graded, but akin to that of everyday dress and civil dress (Jencks 2005).

Cultures and places are constantly constructed and negotiated through the process of constructions of sacredness and powers, of morality and transgression, of nations and identities, of heritage and tourism and many others. Many researchers (as examples see Duncan 2004, Cresswell 1992, W.J. Mitchell 2005) emphasize the role of landscapes in the constructions of identities and meanings, highlighting in the process the ways in which landscapes are both constituted by, and legitimate social power relations. At the same time, attention is also paid to the negotiated nature of meanings, and the roles of landscapes in such negotiations, often employing the notions of hegemony and resistance (Kong 2007). Landscapes do not simply fulfil obvious, mundane functional requirements, nor do they simply represent localised cultural creations, like house styles or barn types. Rather, through the vocabulary of various conventional forms – signs, symbols, icons and specialized tropes in the landscape – people, particularly powerful people, tell morally charged stories about themselves, the social relations within their community, and their relations to the divine and an order (Duncan 2004).

There is no landscape, architecture, design, or sign silent or free of meanings. Buildings or landscapes, as Ruskin noticed, do not only have to look good, but also speak well (Jencks 2005). In some cases the function and role of the landscape is most clearly pronounced. The concept of *architecture parlante* or 'speaking architecture' relays on messages 'manifest' by forms of the buildings or settings, which should explain their own function or identity.[3] In most other cases, value of a sign or a construction relays on its place in broader relation or discourse, while semiotics helps to analyze and understand the process of production of meanings. Meaning is constantly being produced and exchanged in every personal and social interaction in which we take part. Members of the same culture share the same set of concepts, ideas and images which enable them to think and feel about the world, and thus to interpret the world, in roughly similar ways. Sense of place or landscape is a dialogue – always only partially understood, always an unequally exchanged (Hall 2002). Meaning is transmitted and attached to a variety of elements, like sounds, words, expressions, gestures, but also places, buildings and urban structures. Those components don't have any clear meaning in themselves. They are vehicles or media which carry meanings. Those signs and symbols stand for concepts, ideas and feelings in such a way as to enable others to 'read', decode or interpret their meaning in similar way as we do. Every language, as well as landscape, is a signifying practice. It is through culture and language the production and circulation of meaning takes place. Since the cultural turn in social sciences, meaning is thought to be produced or constructed rather then simply 'found' (Hall 2002, Johnson et al. 2004).

Umberto Eco (1985) classifies codes of meanings into three main categories, related to three main attitudes of landscape studies: typological or social, connected with its function; spatial, associated with shape, syntactic, structure of construction; and semantic, which denote the primary functions. Furthermore, Eco (1985) categorizes the modes of sign production using four basic parameters, including physical labour needed to produce expressions, relation between the abstract type of the expression and its tokens, type of continuum or material substance to be shaped in order to produce physically an expression and complexity of the articulation (see also Blonsky 1995, Eco 1986).

Urban form and morphological aspects of urbanization and civilization are created as result of objectification of systems of values and cooperation with members of local societies. Existing urban and architectural structures are being enriched by meanings and connotations. The subtexts define the activities towards and within the urban structures. Meanings are being constantly elaborated in the process of persistent interaction between creators and recipients of urban space. Meanings also create inter-subjective, cultural urban space, which is continuously modified by local societies, according to their current needs and possibilities. Cities, their morphology, functions and identities are resultant of complex net of social relations, which is spread between many layers of urban life, like economy, politics, history, nationality and culture, in addition to relations between them in a given time and place (Krajewski 2002, Hall 2002). Significance of cultural landscape sometimes is quite obvious, but sometimes is deeply hidden behind built forms of functions. There are two main interpretations of cultural landscape, used in new cultural geographies: textuality and iconography. Both of the techniques are anchored in post-strucuralism, positionism and social construction. Text and icon relate mental

---

3 The phrase was propagated by two 18th century French architects: Étienne-Louis Boullée and Claude Nicolas Ledoux.

values to the outward appearance and can be simultaneously used to understand meanings of cultural landscape.

Textual metaphor

Textual analyses of landscape are closely related to the linguistic and semiotic studies and refer to the attributes that distinguish the text or any communicative content under analysis, as an object of study in those fields. It is associated in both fields with structuralism and post-structuralism. The word 'text' arose within structuralism as a replacement for the older idea in literary criticism of the 'work', which is always complete and deliberately authored. A text must necessarily be thought of as incomplete, indeed as missing something crucial that provides the mechanics of understanding. The text is always partially hidden; one word for the hidden part in literary theory is the 'subtext'. The concept of the 'text' in structuralism requires a relatively simple relationship between language and writing (Johnson et al. 2004, Winchester, Kong and Dunn 2003).

There is a long tradition in cultural geography of 'reading the landscape'. The metaphor of seeing landscape as text drew upon the influential work of the cultural anthropologist Clifford Geertz (1973), who suggested that landscapes could be read as a social document, using techniques and methodologies of literary theory. The textual metaphor can be used to illuminate the crucial relationships between landscape and ideology, by helping to identify how landscapes can transform ideologies into a concrete, visual form. Landscapes serve to naturalize asymmetrical power relations and cultural codes. The reading of urban text is aimed to penetrate 'the layers of ideological sediment', recorded and coded in the city forms and structures (Black 2003). The discursive process of researching the landscape as text and relating 'text' to its 'context' is essential to read the multi-layer content of the urban setting. Landscape always represents and symbolizes the relationship of power over which it has emerged and the human processes that have transformed it. The complexity of images written into the city can be interpreted by 'poly-visual' explanation or various reading of urban texts.

Accepting landscapes as texts, broadly defined, we are led to examine how landscapes encode information. This is connected with the concept of intertexuality, which implies that the context of any text is other text (Duncan 2004). In the case of landscapes, the contexts in which they are produced and read may be texts written in other media. Intertextuality broadens the textual metaphor beyond the interaction between different texts and between different types of texts such as written and landscape texts, but also between these texts and social practices which have been textualized. The notion of culture as a signifying system provides a useful framework to examine this transformation of ideas from one type of medium to another. It also makes use of the concepts of discursive field of Foucault (1986). The landscape study probes the role of landscape in the construction of social and political practice. In order to holistically analyze cultural landscape, one must go beyond a consideration of formal semiotic or tropological properties of the landscapes as a system of communication, to see the landscape in relation to both structured political practices and individual intentions (Duncan 1990).

Iconography of landscape

Iconography is based on studies of identification, description and the interpretation of the content of images. In anthropology, sociology, media studies and cultural studies, iconography refers the study of images or signs with an important significance to a particular culture. Discussing imagery as iconography in this way implies a critical reading of imagery that often attempts to explore social and cultural values. The iconography or iconology share certain aspects both with an iconic object, such as a Byzantine painting of Jesus, and the philosophical definition of an icon, that is, a sign with some factor common with the thing it represents. An icon (*eikon*) is literally a 'likeness, image, or similitude', or 'image writing', and comes from the Greek.[4] The word still carries this old religious meaning, as does its antithesis, iconoclasm – the actual destruction of such objects or images. Landscape iconoclasm has followed many conflicts and wars because destroying important symbols is aimed at breaking the spirit and changing representations, but the destroyed icon are not images of 'our', but 'their' gods[5] (Jencks 2005).

Iconographical interpretation of landscape, as highlighted by Cosgrove (1998) gives specific attention to the development of the study of landscape as a way of seeing or representing the world. The aesthetic view of landscape was explained as a way of conceptualizing and signifying the culture. Iconography, as a theoretical and historical study of symbolic imagery, is an interdisciplinary examination, absorbing arts, architecture, anthropology, literature and, last but not least, cultural geography. The landscape idea represents a way of seeing in which people have 'represented to themselves and to others the world about them and their relationship with it, and through which they have commented on social relations' (Cosgrove 1984, 1). Iconography or iconology, according to Gottmann, resists movements and partitions the space. This force, more abstract then material, is based on identity and symbolic links (Gottmann and Harper 1990). Through their iconography, groups share the same representations, visions of the world and values, uniting them within common space of belief. Iconography creates stable identities and helps to maintain these identities by resisting generalized circulation and by partitioning the space. Icons carry a meaning, which they bestow on those places where they provide roots to people. Icons offer an image of the world as much as they make of the individual self in the world: they are a worldview from a particular standpoint (Bonnemaison 2005). Iconography is both expression and matrix of a people's vision of the world; it can be also a way to contextualize cultural landscape.

Iconic landscapes accompany civilization from its very beginning, and they always speak of the times, mark the epistemological character of the place, society and epoch. There is a long tradition of iconic landscape features, like the seven wonders of the world: the pyramids or the Colossus of Rhodes, gothic cathedrals or

---

4 The other meaning of iconography, most popular in the history of art, is the painting of icons in the Byzantine and Orthodox tradition.

5 One of the recent and most spectacular examples of iconoclasm was the destruction of the World Trade Center towers in New York in 2001, seen as an icon of Western capitalism for Al Qaeda (Jencks 2005).

baroque palaces. Recently, thanks to new technologies and advanced engineering, new icons have almost become freed from traditional architectural conventions and limitations, to become more objects of art and sculptures than buildings. The new icons represent globalized societies, so they become global icons. Every epoch has its own 'cathedrals', rising above the ordinary and everyday life and practices: train stations and factories in industrial times, office and residential towers in the mid 20ᵗʰ century, and according to the cliché, museums become the new cathedrals of the age. It shows how far the notion of an icon has travelled, starting life in the Christian past as an object of religious veneration to now become an object of shopping (Jencks 2005). It might also reflect the shift of sacrum in modern societies, where churches become small, modest and human scale in architecture, while shopping malls, art galleries and banks create outstanding landmarks that dominate urban landscape.

Text and icon interact in a stimulating combination that relates to the post-modern 'world as an arena' metaphor. And both of them can be decoded using the same sources and techniques. The text attached to the landscape is usually politically and ideologically declared and written in media publications and other official materials, where the intentions of the decision-makers and designers are being clarified and explained. Sometimes those *landscape lords* articulate the significance of landscape features at various forums, conferences and meetings. The ideas as *signified* features are being visualized in the form of *signifiers*. Architects usually play the roles of the coders, who transfer the idea or text into landscape form and function. The other part of the landscape interpretation is the usually a bit more difficult to deduce: it is the way people, the receivers, see, understand, translate and interpret the landscape. There is no one single mode of interpretation, since every human, and every cultural group can have individual, culturally conditioned ways of interpreting signs and symbols of the landscape. Decoding meanings of cultural landscape can be based on media, when public interacts the landscape, but mainly can be analyzed by interviews, conversations, and recently Internet discussion lists. Additional and very important sources of landscape interpretations are books, especially memoirs, essays, travel books and impressions, where opinions on landscape and its significance is written, sometimes between the lines.

Cultural landscape, as social construction, is a form of spatial and cultural negotiation between representation of the past and imagination of the future. The past is mainly facilitated by histories and memories, whereas the future is conditioned by contemporary powers. Interpretations of history, together with past and present depictions of power, are an integral part of landscape discourse, especially in post-traumatic landscapes.

Chapter 2

Representations of Memories and Powers: Discursive Historical Landscapes

Cultural landscape always mediates between past and future, representing both negative and positive aspects of history, as well as contemporary powers and visions. Authorities, hopes and expectations can be represented on many different levels and arrangements, all of them deeply anchored in local, regional and national historical discourses. Cultural landscape materializes and denotes current conditions as development paths. Metaphor, as was pointed out by Aristotle, is the transport to one thing of a name which designates another (de Certeau 1985). Every cultural landscape is metaphoric in some sense, and conveys memories, histories, experiences, as well as the wishes, needs and structures of authorities. Landscape appears at the historical moment which also sees the development of the theory of linear perspective, this being the technical correlate to the estranged view: both conventions – landscape and perspective – 'reinforce ideas of individualism, subjective control of an objective environment, and the separation of personal experience from the flux of collective historical experience' (Cosgrove 1984, 27).

The idea and practice of landscape is embedded within a system of power relations and suggests that it is exercised as a form of cultural power. Cultural landscape clearly intersects with senses of community, group belonging, history, power and identity, and so in concrete political situations may be drawn upon in the face of specific external threats (Cosgrove 1984). Historical and contemporary powers are interwoven into our representational system, which exemplifies and visualizes the form, functions and meanings of cultural landscape. Lewis (1979, 12) noticed that many contemporary societies lack knowledge of their cultural past. It is this forgetting, 'cultural amnesia', which allows the landscape to act as such a powerful ideological tool. By becoming part of the everyday, the taken-for-granted, the objective, and the natural, landscape masks the artifice and ideological nature of its form and content. Its history as social construction is unexamined, while powers hold and often manipulate national and regional memories. Local accounts of the nature and importance of a given landscape, while situated within and structured by a general cultural discursive field, can at times differ sharply either within or between groups. Such discursive spaces, or 'openings' to use de Certeau's (1985) term, could prove to be one of the most fruitful areas of research into the signification of landscapes (Duncan 1990). Every cultural landscape is in a sense a discursive historical landscape, which reflects memories facilitated or enhanced by authorities, as a consequence of representation system.

Landscape as representation system

For many postmodernist researchers (see Söderström 2005), representation is at the core of social scientific practice and is seen as summarizing the process of knowledge production. For Heidegger the word becomes a conceived image (*Bild*), while for Foucault representation remains an epistemological model in the Modern Age (see Heidegger, Fried and Polt 2001, Foucault and Gordon 1981). The Representational Theory of Mind (see Fodor 1981) postulates the actual existence of sort of mental intermediaries between the observing subject and the objects, processes or other entities observed in the external world. These intermediaries stand for or represent to the mind the objects of that world. The *image* has lots of similarities with the notion of cultural landscape, and can be correspondingly understood and explained as an essential and visual part of the process of representation.

Many authors, like Fodor (1981) and Pinker (1991) believe that representational systems consist of an internal language of thought. The contents of thoughts are represented in symbolic structures which, analogously to natural languages but on a much more abstract level, possess a syntax and semantics very much like those of natural languages. Fodor's (1981) language of thought hypothesis states that cognition is a process of computation over compositional mental representations. This means that thoughts are represented in a language, sometimes known as *mentalese*, which allows complex thoughts to be built up by combining simpler thoughts in various ways. Mental states, such as beliefs and desires, are relations between individuals and mental representations. These representations can only be correctly explained in terms of a language of thought in the mind. Fodor (1981) adheres to a type of functionalism, maintaining that thinking and other mental processes consist primarily of computations operating on the syntax of the representations that make up the language of thought.

Representation is strongly connected to the process of spatialization. Space is equivalent to representation, which in turn is equivalent to ideological closure (Laclau 1990). Massey (2006) points out that space conquers time by being set up as the representation of history/life/the real world, while spatial order obliterates temporal dislocation. Spatialization is often equivalent to hegemonization: the production of an ideological closure, a picture of the essentiality dislocated world as somehow coherent. 'Any representation of a dislocation involves its spatialization' and the same time dislocation destroys all space and, as a result, the very possibility of representation (Laclau 1990, 72). The relation between time and space is an important part of social science discourse. Laclau (1990) writes about hegemonization of time by space through repetition, while for de Certeau (1985) the 'proper' is a victory of space over time, of 'representation' over 'reality', of stabilization over life, where space is equated with representation and stabilization. There is a crisis of representation in sense that that it must be recognized as constitutive rather then mimetic. Representation now might be no longer process of fixing, but an element in a continuous production (Massey 2006).

For a growing number of researchers (Hall 2002, Gottmann and Harper 1990, Meining 1979, D. Mitchell 2005) landscape, in similarity with language, can operate as a representational system. Signs, names, buildings, places and spaces can be read and interpreted as geosymbols. Landscape is one of the most visible and communicative media through which thoughts, ideas and feelings, as well as powers and social constructions are represented in a space. Representations through landscapes are therefore central to the process by which the meaning of space is produced. Members of the same culture share same values and meanings and must reveal the same or similar systems of communication, based on mutually understood codes and signs. Two main approaches to representation work: semiotic, concerned mostly with how language and landscape produce meanings; and discursive, more focused on how effects and consequences of representation can simultaneously coexist and create a base for further interpretations. Cultural urban landscape is a system of representation, by which all sorts of objects, buildings, features, people and events are correlated with a set of concepts or mental representation we carry in our heads (Ashworth 1998). The system consists of 'individual concepts, but of different ways of arranging, organizing, clustering and classifying concept, and of establishing complex relations between them' (Hall 2002, 17). Since culture is the core of any meaning-making process, there are two 'systems of representations'. One enables us to give meanings to the world by constructing a set of correspondence between things, like objects, events, people and places, and our system of concepts. The second one depends on constructing a series of associations between our conceptual map and a set of signs, organized into various languages which stand for those concepts. The relation between 'things', concepts and signs lays in the heart of the production of meaning, while the process which links these three elements together is called representation (Hall 2002, 17–19).

Landscape remains a representation of what is and what can be. The landscape has enormous inertia, an inertia made real not only in bricks and stone, but also in people's livelihoods and homes (Harvey 1982). Space, as Lefebvre (1991, 143) says more generally, is not produced 'in order to be read and grasped, but rather to be lived by people with bodies and lives in their particular (…) context'. Landscapes are read and struggled over because the meaning attached to the landscapes, working together with the landscape's built form, establish the conditions of possibility for people live in that place (D. Mitchell 2005, 51). The landscape, as a vast, humanly created but fixed and not easily destroyed reservoir of use-value and as a physical and ideological representation of what is and what is not possible in any given time, of what is right, just and natural, is both an outcome of struggle and a mediator of it. The importance of landscape derives from how a morphology, an arrangement or way of seeing work as combination and become the vehicle of all manner of exclusionary, alienating, expropriating social practices. 'To see the power at work in the landscape requires attention not just to the landscape (as a form, representation or set of meanings) in and of itself, but to the social relations that give rise to and make possible landscape's ability to do work – to function as a reification and a fetishization – in modern societies' (D. Mitchell 2005, 54).

In a study of how people see the city, the urbanist Kevin Lynch (1960) has asserted how important it is to concentrate on the particular visual quality. The apparent clarity of legibility of the urban landscape plays a crucial role in the process of representation. Lynch reported that users understood their surroundings in consistent and predictable ways, forming mental maps with five elements: paths, formed by the streets, sidewalks, trails, and other channels in which people travel; edges, perceived boundaries such as walls, buildings, and shorelines; districts, relatively large sections of the city distinguished by some identity or character; nodes, focal points, intersections or loci; and landmarks, readily identifiable objects which serve as reference points. Those elements can be also vital elements of cultural landscape, especially paths, nodes and landmarks.

The significance of the outward appearance of landscape is based on the process of representation, which is composed of two procedures. The first enables us to give meaning to the world by constructing a set of correspondences or a chain of equivalences between people, objects, events, abstract ideas, places, and so on, and our system of conceptualization. The conceptual map is the result of the process of giving meaning. The second system depends on constructing a correspondence between the conceptual map and a set of things, arranged or organized into various languages that stand for or represent those concepts. The relation between things, concepts and signs lies at the heart of the production of meaning of urban landscape (see Hall 2002). Theory of representation can be most useful in interpreting cultural landscape, especially using textual or iconographical metaphors. Since landscape can be seen as a form of communication or language, with buildings, places and objects as signs or words, the scenery of surrounding is a result of both, mentioned above, elements of system of representation. The production of meaning relates places, events and features of urban scenery with our system of concepts, while representation associates our conceptual map with a set of signs or geosymbols. There are broadly speaking three approaches to explain how representation of meaning works through language and landscape (Hall 2002):

- Reflective or mimetic approach, where meaning is thought to lay in the object, place, event, and building in the real world, and language functions like a mirror to reflect or imitate the true meaning as it already exists. There are some landscape features, like monuments of names, where the meaning seems to be truly located within the object. The sculpture or street name more or less truly reflexes the icon, and many people believe that landscape mimes a system of concepts.

- Intentional, based on belief, that signs and symbols mean what the author intends they should mean. Communication depends on shared conventions and codes, therefore to be understood one has to enter into the rules, ciphers, standards and practices of the social/cultural group. The intentions of the investor, designer or decision-maker can be sometimes clearly read in urban text, but sometimes the initial purpose is forgotten or erased by new *intentioners*.

- Constructivist attitude emphasizes the social character of language/landscape and relies on presumption that things don't mean anything on their own; people do construct meanings, using systems of representation, their concepts and signs. The meanings are not conveyed by any feature of the material world, but by which system we use to represent our concepts. It must be remembered that one object or feature can have different constructivist meaning, dedicated by different social groups. Frequently groups of youngsters use systems of representation distinct from the older generation, so places and urban features might have separate constructivist meanings for them.

As a representation, landscape is also an ideology. The concept of landscape as ideology has its roots in Renaissance Italy and, to some extent, Flanders. It developed as a means of representing a certain relationship between landowners and the land during transition from feudalism to capitalism in Europe (Cosgrove 1984). Ideological landscape represents a way in which certain classes of people have signified themselves and their world through their imagined relationship to nature, and through which they underlined and communicated their own social role and that of others. For centuries the authorities have made many efforts to represent strengths, capacities and structures of powers in visible and stable forms, functions and significance of cultural landscapes.

Landscapes of powers/powers over landscapes

Power, as one of the most important concept of social sciences, is usually defined as relation between two groups or individuals, where one of the group or individual can influence or/and control behaviour of the other one. Michel Foucault (1975) analyzed the structures of power and instead of focusing on localizable, dominant, repressive, legal centres, he turned it to bear on technical machinery and procedures, those 'minor instrumentalities', that, through a mere organization of 'details', can transform 'diverseness' of humanity into a 'disciplined' society, and manage, differentiate, classify and fit into a hierarchy every deviancy that can affect training, health, justice, and the army of labour (Foucault 1975). 'The tiny plots of discipline', the 'minor but flawless' machinery that colonized and made uniform the institutions of the state, derive their effectiveness from a relationship between procedures and the space they redistribute to create an 'operator'. They set up an 'analytic arrangement of space' (de Certeau 1985, 128) and produce a landscape of power which is a cultural artefact of its time and place. The representations of ideas and powers are framed by a contract between strategies and tactics (de Certeau 1985). A strategy is defined as relating to an already constructed place, static, given, a structure. Tactics are the practices of daily life that engage with that structure. The power in society is materialized as a monolithic structure or static strategies, interacting with the everyday tactics of the weak (Massey 2006).

Mechanisms and procedures of reproduction of social hierarchies create a basis for a symbolic economy (see Bordieu et al. 1999). In opposition to Marxist analyses, Bourdieu criticized the primacy given to the economic factors, and stressed that the capacity of social actors to actively impose and engage their cultural productions and symbolic systems plays an essential role in the reproduction of social structures of domination. Bourdieu sees symbolic capital – any species of capital that is perceived through socially inculcated classificatory schemes – as a crucial source of power. When a holder of symbolic capital uses the power this confers against an agent who holds less, and seeks thereby to alter their actions, they exercise symbolic violence. People come to experience symbolic power and systems of meaning as legitimate.

Symbolic violence as the capacity to ensure that the arbitrariness of the social order is ignored and thus to ensure the legitimacy of social structures plays an essential part in this sociological analysis. Symbolic violence is in some senses much more powerful than physical violence in that it is embedded in the very modes of action and structures of cognition of individuals, and imposes the vision of the legitimacy of the social order. Bourdieu focuses on the ways in which power is exercised through the manipulation of symbolic economy (Bordieu et al. 1999). Cultural landscape is one of the most stable, often par excellence concrete, and usually clearly readable and interpreted exemplifications of symbolic capital (see Figure 2.1).

Figure 2.1 Icon of Soviet power in the heart of the city:
Palace of Culture and Science, Warsaw, 2004

Cultural landscape is a political project. This unique composition or palimpsest, representing and reconstructing relationships of powers and history through the system of signs, is written on many layers, including aesthetic, political, ethic, economic, infrastructural, legal and many others (Cosgrove and Daniels 2004, Black 2003). Since the main features of culture landscape represent social and cultural relations and power (in very broad sense of the word), the study of representation of supremacy becomes one of the most imperative research tasks. Urban scenery reflects power, need and aspiration, as well as a glorious and tragic history, written into the symbols and signs (Zukin 1993). Landscapes are the bodily expressions of the ways of thinking, the experience, and the hierarchies of values and culture of each of the group as well as of each individual (see Robertson and Richards 2003, Cosgrove and Daniels 2004, Winchester, Kong and Dunn 2003).

Landscape mediates both symbolically and materially, between the socio-spatial differentiation of capital implied by the market and the socio-spatial homogeneity of labour suggested by space (Zukin 1993). The territory is the nexus of power. To carve and control places within the territory is to dominate. Control of the land mediates the relation of authority between people and nation (Bonnemaison 2005). Cultural landscape is one of the main representing languages of modern society, which signify the spiritual dimension of the investors, architects and users. The context is central to understand the landscape, as it frames and embodies economic, social and cultural processes. The aesthetic form is never neutral – the power is written into the landscape through the medium of design, usually used and overused by rulers to stress the authority and legacy (Markus and Cameron 2002, Winchester, Kong and Dunn 2003). The way landscapes signify relationship of power denotes sociological and economic condition of local societies. As Clifford Geertz (1973, 448) would phrase it, landscape is 'a story ... [people] tell themselves about themselves'. This narration of supremacies, politics, cultures and economies reflexes how contemporary powers are spatially and temporally contiguous with the previous ones. Cultural landscapes become very audacious physical and cerebral expressions of power relation between one group and another. Power, control and resistance are core foci in the discussion about cultural landscapes, and this discussions and processes are always most active and visible in the times of transformation (see Johnson et al. 2004, Best and Kellner 1991, Czepczyński 2005a).

Power over landscapes is usually visualized in the most picturesque, stable, perceptible and spectacular ways. Ruling over landscape, both forms, function, and particularly meanings, becomes one of the priorities of power, especially of those, whose legitimacy is or can be somehow challenged. The need to show his or her rights, authority, control, as well as supremacy and prerogatives is tremendous and typically materialized in grand cultural landscape projects. Those projects usually expound not only political and economic powers, but also, and often above any other, the cultural dominance of the new leaders. The messages coded in cultural landscape are typically very clearly readable for most of the society, and are frequently enhanced by heavily marketized texts. One of the most important features of landscape is pursuit to total domination over the user/spectator. There are three main types of totalitarian spatial structures: monumental landscapes, designed for grand mass gatherings of the crowds; 'exclusive' space for chosen, although sometimes very numerous, people 'included' in to the system, and the 'elitist' areas for small groups of the people of the power (Massey 2006). Functional system of cultural landscape can be analyzed by processes of significance-making, and creation of meaningful places and spaces.

The basic meanings within cultural landscape seem to be created by three main groups of landscape makers and participants:

- *Investors/decision makers* begin the process of transformation of landscape. Their political and/or economic powers enable them to initiate decision of new shapes of cultural landscape. The form usually follows from the function, which associates the text to a particular landscape feature. Investors and *landscape lords* are mainly responsible for function, although sometimes also interfere with the form and can only suggest a future meaning of the feature.
- *Architects/designers* implement the investors' visions and expectations and turn them into brick/concrete/steel solid features. Usually they are responsible for the shape and visual aspects of a feature. Architects follow fashions and trends and shape the function in material form, coding meaning into detail and appearance.
- *Users/percipients* are the final group of landscape users. Inhabitants, consumers and tourists are probable the most important group in landscape discourse. They verify the intentions of investors and designers' aesthetic creations by everyday practice of decoding and re-coding of cultural landscape. Their interpretations are most crucial for the success of the landscape project, since they signify and give meaning/social capital to cultural landscape (Czepczyński 2006c).

In Cosgrove and Jackson's (1989) estimation, culture's retheorization should also take into consideration contestations between groups, evident, for example, in the appropriation and transformation of artefacts and significations from the dominant culture by subordinate groups as forms of resistance. Acknowledgement of this relation necessitates the prior recognition of the existence of a plurality of cultures which are time and place specific rather than the hitherto implicitly held assumption of a unitary culture (Kong 2007). The ideas of hegemony and resistance between capital and social classes are reflected in their representations of landscapes. Power written into the visible forms of urban structures was particularly strongly featured in totalitarian regimes, especially communist ones. Most important and magnificent urban designs have always been created for strong, ambitious power or for well-managed and rich societies. Power over and management of ideological and everyday landscapes are extremely important and often-proclaimed tools for specific marketing, image making and solving real social problems.

Form or architecture is a distinctive language of communication between investor and user, filtered by the architect's design. Usually the message is well understood but is sometimes differently interpreted and may cause misunderstanding. The understanding of the message mainly depends on the user, and his or her ability and will to understand and communicate with the sender. Usually the manipulation of the meaning is based on the manipulation of the user, and people are often dangerously open for given or 'granted' interpretations. The universal mimetic transcription of signs seems to be merely an idealistic fantasy (Massey 2006).

Every ideology influences, censors and manipulates cultural landscape. The most obvious manipulations include forms, texts and spatial organization. Scale or size indicates the significance and importance, forms connotes historical memory and imaginations of the past, organization of space can increase or diminish certain social groups. Manipulation of each of the mentioned elements leads to creation certain social constructions, ideologically correct and expected. Despite of all the manipulation techniques and procedures, the final result, measured by human perception might be slightly or completely contrary to the initial intentions of the decision powers (see Chapter 3, Non-socialist features of socialist cities). Cultural landscape has always been open and full of interpretations, so full control over cultural landscape is not possible: one can control form or function, but can never entirely control the interpretations of landscapes, which are the core of its communication. Landscapes of power are always very changeable, unstable and subjectively judged, as any political power can be. Powers, their significance, and landscaping techniques and practices are being constantly negotiated and modified. Since landscape is, in a sense, the state of a culture, petrified in any give moment and place, it is also very closely related to history. Learning, memory and remembrance play the most influential role in cultural landscape making. Historical policy seems to play a more and more important role in many, especially post-socialist, countries, like recently in Poland and Hungary, where many unanswered historical questions or pains return as political issues and arguments (Massey 2006).

History, memory and oblivion

If cultures, places and landscapes are socially constructed, so too is the past. Socially produced and constructed cultural landscapes, as much as any other political statement, can be seen as 'centres of human meaning as well as mode of social control and repression' (Tilly 1994, 19). The mechanisms of restraint are usually rooted in the past, while the interpretation of the past can be, and frequently is, a political assignment. In the same way that forms of materialized historical interpretations, artificial materializations of the past produce meanings and construct reality. In reference to Foucault (1975), memory, as representation of the past, is an important political resource: 'memory is actually a very important factor in struggle ... if one controls people's memories, one controls their dynamism ... It is vital to have the position of this memory, to control it, administer it, tell it what it must contain' (Foucault 1975, 25–26). For Orwell (see 1949), the past manifested in memory practices of commemoration and rejection influences contemporary identities and, to a further extend, future opportunities and developments. In reference to those two, any identity, also historical identity, is not fixed, but socially constructed (see Graham 1998). Baker (1985, 135) argues that 'the past is seen to influence or even determine the present'. He also points to the fact that the representations of the past tent to minimize diversity and complexity, bestowing 'on past experience as overriding sense of unity'.

Historically conditioned identities and cultural codings do not remain stable; they can and must be continuously reflected upon and negotiated. Different disciplines, among them geography, consider commemoration as a social activity (see Halbwachs and Coser 1992). Furthermore, it is an expression and an actively binding force of group identity. Collective, social, public, historical and cultural commemoration serves to give oneself a coherent identity, a national narration or a place in the world (Said 2000). Commemoration is a part of historical policy, which finds its materializing form in cultural landscape features. Landscapes contain the traces of past activities, and people select the stories they tell, the memories and histories they evoke and the interpretative narratives they weave, to facilitate their activities in the present and future. The process of selection of memories is condition or determined by several factors, most of which related to the past or circumstances.

For Ortega y Gasset (1996) individuals and societies are never detached from their past. In order to understand a social or cultural reality we must understand its history. In Ortega's opinion, history and reason should not focus on what is static but on what becomes dynamic. Ortega y Gasset (1996) suggests that there is no me without things and things are nothing without me, I as a human being cannot be detached from my circumstances or my world. This led Ortega to pronounce 'I am myself and my circumstance'[1] which stresses the role of external factors facilitating personal development and possibilities and proposes a system where life is the sum of the ego and circumstance, both historical and geographical. This *circunstancia* is oppressive; therefore, there is a continual dialectical exchange of forces between the person and his or her circumstances and, as a result, life is a drama that exists between necessity and freedom. Fate, circumstance and history of humanity can be by definition seen as path-dependent (see more David 2001, Garrouste and Ioannides 2001, Mach 2006). Path dependence has primarily been used in comparative-historical investigations to analyze the development and persistence of institutions, whether they are social, political, or cultural. In the critical framework, antecedent conditions define and delimit agency during a critical juncture in which actors make contingent choices that set a specific trajectory of institutional development and consolidation that is difficult to reverse. Institutions, as well as states and politics, are considered by most of the scholars to be *path-dependant*, as they constrain choice to a limited range of possible alternatives, reducing the probability of path changing and presenting an evolutionary tendency, given the acquired routines. Agency takes place within given context and path dependency is one of the most likely, but not the only, results of the interaction between the two, which brings about relative stability (Kazepov 2005). The stability is visualized and infixed in material and mental features of cultural landscape, facilitated by political and economic powers.

In many societies, especially in nation states and other passionate political organization and parties, one dominant historical narration and memory dominates over any other possible interpretations of the past. Typical of the 19th and 20th century, one meta-narration can be, and recently often is, replaced by a *polyphonic memory*, consisting of few corresponding and supplementary interpretations and memory traditions (Traba 2006). This concept of *lieux mémoire* or 'memory places' has been

---

1 Spanish '*Yo soy yo y mi circunstancia*'.

introduced by French historian Pierre Nora (2006), who suggests a new form of historical writing and interpretation, what he calls 'history of second degree'. Linear and neo-positivistic factography is being replaced by scores of symbolic spaces and landscapes. Those *lieux mémoire* are not only constantly present in social memory, but also facilitate and enhance local and regional identity and consciousness. The concept of *lieux mémoire* is reflected in many cultural interpretations of landscape where memory is a significant part of culture, typical and representative for a chosen social group. Memory, either collective or individual, does not seem to be the true record of past events, but a kind of text which is worked upon in the creation of meaning. Identities are continually crafted and re-crafted out of memory, rather than being fixed by the 'real' course of past events (Thomas 1996). The memory is often treated by many modernist historians, social scientists and politicians as an untouchable relict of the sacred past, unaware or/and unwilling to realize multiple versions and trajectories of the memories and interpretations of the very same event. Within nation states, history, memory and heritage tell powerful stories, usually the ones that stress stability, root, belongings and heroism, but those stories always naturalize a particular sort of social relations and are often are far from the objective, historical truth (Bender 2001). Power over historical memory can be important tool of historical policy, used to legitimate present actions. Orwell (1949) summarized the role of historical policy, pointing out that he who controls the past commands the future; he who commands the future controls the past. This statement can be exemplified by many cases of totalitarian and post-totalitarian landscapes re-interpretations. There is a common historical tradition and tendency to discuss the supremacy of one track of memory over the other, but recently many researchers try differentiate between the tracks and understand them (Massey 2006).

History is being transferred and transformed by memory practices. There are two main traditions in memory research: one is focused on 'who' remembers, the other is more concern on 'what' is to be remembered. Both of the attitudes are closely connected and are essential to understand and interpret process of memorising and recalling of the past (Ricoeur 2004, 3–4). There were two processes of memorizing, related to two words in ancient Greek – *mnēmē* and *anamnēsis* – to mark, on one hand, a memory as something that comes passively, unintentionally, while the other word stands for recollection as object of searching, remembering and recalling. Remembering, then, means both having a memory and searching for it. This cognitive and pragmatic dualism is reflected in memory claim to be 'true', which has to be confronted with the historical and more objective sources (Ricoeur 2004). Historical policy and its landscape representations reflects the process of *anamnēsis*, where landscape features are being searched, remembered and related to past events.

The process of memorizing can be also described as opposition between two main courses, one *keeping memory in mind*, and the other *pursing memory beyond mind* (Casey 1987). Those main courses are being implemented by mnemonic modes: reminding, reminiscing and recognizing. Reminding is based on signs which are supposed to protect from forgetting. Reminiscence is far more active than reminding and is established by enlivening the past by its multiple evoking and common strengthening of shared events or collective knowledge, where one's memories are reminders for the other's. The third type of memory, recognizing, is connected with the presence of something we met before, but which is absent now. The 'thing' is doubly different: as absent, different then now and past, different then now existing. 'Present is being overcastted by otherness of the past. This is why a memory becomes representation in double sense of prefix 're-', as something backwards and new' (Ricoeur 2004, 38). Reminiscence is not simply based on evoking the past, but more on a realization of the learnt knowledge, deposited in mental space. This memory is exercised, nursed, trained and created. There is an ethical-political level of memory: an obligated memory. Command to recall can be understood as an encouragement to simplify history. Memory 'can be seen as a matrix of history, but can be also used as demand and stipulation of memory against history' (Ricoeur 2004, 86).

Halbwachs connected memories directly to the social group or society (Halbwachs and Coser 1992). Reminding is strictly connected with acceptance of one or many social groups and this point of view is included into one or many streams of collective thinking. Personal memory is often facilitated by a much more stable collective memory, enhanced by media, education and print materials. Both personal and collective memories can be manipulated for various, mainly political reasons. Human memories or retrospectives are always enhanced by verbal, visual or formal stimulations. 'Even the landscapes that we suppose to be most free of our culture may turn out, on closer inspection, to be its product' (Schama 1995, 9). Landscape is read and appreciated through the cultural and historical memory the people bring to it. 'Vast and seemingly impersonal historical and/or economic 'forces' have always been the aggregate products of the choices that were made by individuals' (Rykwert 2000, 9). Each social group have constructed its cultural landscape out of specific tracks of memories or *anamnēsis*, which becomes *believabilia* and *memorabilia* (de Certeau 1985). Facts and events we remember, recall and believe develop into meanings and significance of cultural landscape features.

Riesman (with Glazer and Denney 2001) suggests that the *inner-directed* man, conditioned by inner rules and own morality, has been recently replaced by the *other-directed* man, conditioned by reactions of other people, when their expectations create standards of behaviour. The *other-directed* man often doesn't really know what he or she wants, but clearly knows what he/she likes. The other category of 'round of life' includes *tradition-oriented* man and society, determined by history and past patterns of manners and activities. Memorial policy plays an important role in many governmental systems, especially the ones dominated by *tradition-oriented* man, dominated by history and past (Riesman, Glazer and Denney 2001). 'Historical justice' discourse becomes one of the major projects, while social and national memory is used to legitimate and strengthen the political power. The *tradition-oriented* powers try to institutionalize memories to control the interpretations of the past. Presently, many right wing governments of Europe seem to be *tradition-oriented* to some extent. In some countries, the control over traumatic and often adulterated memories is being facilitated by special historical institutions, established to explain, interpret and disseminate real/preferred/factual/chosen/ favoured history.[2] The institutions like the Institute of National Remembrance – Commission of the Prosecution of Crimes against the Polish Nation (*Instrytut*

---

2 Tick as appropriate.

Pamięci Narodowej) or the Institute for the Investigation of Communist Crimes in Romania (*Institutului de Investigare a Crimelor Comunismului*) were established to clarify and adjust interpretations of the ambiguous communist period. It seems that many post-traumatic and 'past-oriented' societies become very much *tradition-oriented*, dominated and sometimes overshadowed by past.

Memory is being 'archivized' in minds, what often is very changeable and unstable, but also in written forms, as well as in material artefacts, like landscape features. Both burdens and glories of history have their landscape representations. The cultural landscape can be analyzed as an icon of memory, but we must remember the weakness and human character of memory, especially its selective process of recalling. Materialized and institutionalized features of memories become authorized elements of memorial policy, sometimes, especially in authoritarian regimes, aimed at abusively controlling memory (see Orwell 1949). These features include teaching programmes, lists of historically important buildings, publications and celebrations of historical events, all of them based in landscape settings, where both forms and meanings of landscape play an important, sometimes crucial role. Places of memory, commemoration, forgiveness, pride, dignity, shame, infamy and blame create mental maps of every society, where treasured *sacrum* often neighbours dishonourable *profanum*. This place–memory discourse becomes more noteworthy in transitional societies, when changing political and social system implies changing reminiscence and recollection of the past. *Pre-* and *post-* landscapes have always played important role in featuring societies in transition. 20th century had witnessed numerous major shifts of national and regional development trajectories, followed by key memory and landscape reinterpretations. *Pre-* and *post-* revolutionary, capitalist, colonial, fascist, socialist landscapes have been reconstructed due to reinterpretation of past, based on control of reminiscence and memories. The most popular way to deal with redundant memories is to recall only the features of the past which go well with the present system, and eliminate any which work against the new structure (see Chapter 4).

Nation or society can be analyzed as a community connected by memories and obliviousness (Renan 1995). Every community needs some emotional binders incorporated into its institutions, symbols and narrations. Interpretations of the past are always politically conditioned, and they often become political battlefields. Cultural landscape anchors national, regional and local traditions of patriotism and commemoration, particularly during periods of political change. Each nation and social group has developed traditions and rituals designed to define and solidify their sense of group identity. These traditions revolve around group heroes, victories and accomplishments, and they come to be celebrated in literature, poetry, historical writing, opera, popular music, theatre, painting, sculpture, architecture and landscape design (Foote, Tóth and Arvay 2000). Cultural and historical geographers have recently turned to examinations how commemorative traditions emerge in landscape and in built forms (see Cosgrove and Daniels 2004, Atkins at al 1998, Tilly 1994). Secular efforts at tradition building mirrored, at least in function, court rituals and religious traditions that had evolved in Europe over centuries. 'These traditions became a means of legitimising political and territorial claims over increasingly large and diverse populations and for coming to terms with the social and economic upheavals brought about by industrialization' (Foote, Tóth and Arvay 2000, 304). Many also face disjunctures, breaks, and gaps that must be ignored or bridged. This has meant assembling commemorative traditions from a variety of national, religious and ethnic sources and linking together events from many different centuries (Foote, Tóth and Arvay 2000).

Memory is materialized and hardened in the forms and meanings of heritage. History and heritage – what we opt to select from the past – are used everywhere to shape emblematic place identities and support particular political ideologies (Graham 1998). There is a need to relocate oneself in the given network, the search for rediscovery of the cultural self and my own cultural tribe or group. What to keep and what not to keep is an indicator of social aspirations and desired cultural identities. This re-formulation is bound for both local societies, as well as toward investors and tourists to show both where are we coming from, or rather – where we would love to see us coming from, and where we are going. This selection of particular historical periods, their mystification and political use is typical of many transitional societies. We generally clearly know what we do not want, and look for the references in the past. Cultural and political history of nations, societies and cities have been constantly negotiated in landscapes as identities, based on what is remembered or rather recalled.

Heritage and cultural landscape

Cultural heritage (or just heritage, since every heritage is cultural), is the legacy of physical and mental artefacts and intangible attributes of a group or society that are inherited from past generations, maintained in the present and bestowed for the benefit of future generations. Often though, what is considered cultural heritage by one generation may be rejected by the next generation, only to be revived by a succeeding generation. A broader definition includes intangible aspects of a particular culture, often maintained by social customs during a specific period in history. Heritage is reflected in ways and means of behaviour in a society, and often formal rules for operating in a particular cultural climate. These include social values and traditions, customs and practices, aesthetic and spiritual beliefs, artistic expression, language and other aspects of human activity. The significance of physical artefacts can be interpreted against the backdrop of socioeconomic, political, ethnic, religious and philosophical values of a particular group of people (Czepczyński 2004, Schröder-Esch and Ulbricht 2006).

Heritage landscape is always socially constructed, whether by intent or default. Identification of cultural heritage is crucial in order to analyze and interpret cultural landscape. Only accepted and assimilated succession becomes real heritage, which is always connected with both culture and history, two multifaceted and challenging concepts. Heritage can be perceived as a stock, received inheritance, but also as bequest or succession. Heritage can also be a selected part of the past, used for contemporary economic, social, cultural or political purposes (Graham and Ashworth 2000). If culture can be seen as the total information generated by the previous generations, cultural heritage is the part of the past that is necessary to sustain a social link with the group and preserve cultural identity. The heritage is the fraction of the bequest, recognized and accepted by the heirs (Kieniewicz 2002). Heritage can also be understood as an external exemplification of culture and the most important of the transmitters of codes, necessary for placing oneself in cultural, historical and social context. Inheritance can be analyzed on a combination of levels, including the personal, institutional, social, regional, national, and global or civilizational. The framework of heritage can be also specified in a range of contexts, including historical, social and spatial. The historical context refers not only to the past of the place, but also to the past of the social group and often to the personal descendants' history. The social background varies according to ethnic and national milieux, but also according to gender, age, social class and family upbringing. The spatial context or sense of place identity seems to provide the perfect compromise and unifying platform for historical and social perspectives. A sense of place based on heritage also shows the importance of space and personal landscape. For many, spatial identity or heritage sense of place can be the easiest way to discover the answer to the question of self-identity.

Landscape often carries strong emotional values, connected with personal, familiar and national connotations. The emotional value of landscape is repeatedly transferred from generation to generation. Personal attitudes and feelings towards cultural landscape are seen as a crucial component of post-modern culture. The inner landscape or the imprinted setting within us, oriented by a personal 'cognitive map', leads us and often determines our spatial behaviour. The personally defined social landscape, based on private experiences, marks out the meaning and significance of the visible urban scenery. The heritage landscape speaks, but not everybody can understand the language of the walls and streets (Czepczyński 2004). The architectural dimension of the landscape is always a visualization of the dreams and desires of the investors and architects, and sometimes of its future users. Some historical objects carry great emotional value for their heirs. The demand to keep the structure as original as possible is often immense, and is usually made for the sake of the future generations. The 'perfect' or 'original' condition is habitually carefully chosen, as the decision-maker picks out selected, particularly glorious moments of history, he or she wants to preserve. In these turbulent and insecure times, when everything changes so fast, stable, old splendour seems like an anchor or lighthouse that stabilizes one in place and time, which reminds one of his or her identity and origin. Looking back to the values of heritage landscape is much safer than looking forward to the uncertain future (see Figure 2.2) (see Ashworth and Turnbridge 1999, Mach 2006).

One of the crucial aspects of heritage is the relation between heritage and its heirs. Personal heritage choices are conditioned by a variety of circumstances, including family background and history, education, occupation, gender, age and place of residence. There is no obligation of choice according to birth, family, religion, nationality or ethnicity, although the above-mentioned limit and strongly influence the decision. Cultural heritage is generally accessible for everybody, though some assimilation procedures are necessary to 'tailor' the available heritage stock to the existing, socially and personally conditioned, demands of heritage. Heritage can be easily lost, forgotten or eliminated; it can also, even after centuries of non-existence, be brought back to life (Kieniewicz 2002). One of the most important factors of the inheritance process is the ability and will to read, decode and incorporate inheritance. The will, possibility and ability of the assimilation of inheritance depend mainly on decoding and accommodating the capacities of the heirs. The sensible selections taken by successors are influenced not only by traditions, national and family codes and education, but also by the general geopolitical situation, employment, and ethical and religious issues (Kieniewicz 2002). The identification with heritage is usually a positive process, connected with the affirmative objects, structures, and feelings that facilitate further development. The received inheritance and codes need recognition and advance affiliation by the chosen social group (Czepczyński 2004).

Figure 2.2 Discursive heritage landscape: Gdańsk historical centre, reconstructed in the 1950s, 2003

Personal feelings and inner values are extremely important in determining the quality of life of the inhabitants. The majority of external factors that influence our emotions come from the surrounding landscape. The quality of the cultural landscape is one of the most important aspects of the quality of everyday life. Living in the heritage landscape is often not an easy task, and often quite risky, but can also be an exclusive privilege and a benefit. The landscape layered by a rich palimpsest of history is both an opportunity and threat for quality living. It seems not always easy to take advantage of the values imprinted onto the heritage landscape without a particular preparation and incorporation of the inherited cultural assets. Finally, the procedure by which cultural landscapes are put to use is always a deeply personal one. Since technically everybody lives in a broadly understood heritage place, even if sometimes the process of inheritance is very short and shallow, the problem of 'heritage living' concerns everybody, and everybody chooses and makes heritage decisions. The rich palimpsest of history looks like an archaeological stratigraph, where the layers of historical events are placed one on top of the other (Borley 2002). Heritage is, after all, what we want to have and preserve from history. The selection criteria are conditioned by many factors that create current demand for heritage, including fashion, literature, film, education, and political, social and economic experiences (Ashworth 2002).

Landscape can be seen as a feature and function of a place as well as a development option. That option requires affords of choice of the heritage (Ashworth 2002). The use of heritage landscape in the process of elevating the quality of life is a challenging and demanding practice. To use heritage landscape as an influential feature of the quality of life, one needs to assimilate and accommodate the heritage and the landscape of the place one lives in. Cultural heritage can be seen as a process by which heirs, as decision-makers, selectively choose the preferred parts of the vast stock left by previous generations. The choices often take an eclectic form, since the heirs are not obliged to choose everything. Heritage implies a polysemic set of meanings. These meanings are being permanently transformed and altered with the passing of generations. The transformation is additionally hastened by political, cultural and economic events, such as revolutions, wars or crises. Good living within a heritage landscape requires a particular style of life, ability and, most importantly, the readiness and will to enjoy the quality of the scenery. The suitable hierarchy of values, knowledge and recognition of the past seems to be essential to take comprehensive advantage of the prospects and possibilities that landscape has to offer. The particular, heritage orientated *genre de vie* has a number of specific characteristics. These include a constellation of attitudes, images and perceptions closely related to the inhabited landscape. It is also closely tied to group and individual identities and constitutes an interactive network between the socio-economic features and personal sensitivity. The heritage landscape is commonly a certain type of a landscape niche; usually exceptionally unique and unrepeatable (see Buttimer 2001, Schröder-Esch 2006, Czepczyński 2004).

The glorious or tragic past, represented by significant objects and symbols, plays a important political and social role. The historical period chosen for restoration and preservation speaks to the social and contextual preferences. The shape of the historical landscape reflects the choices of decision-makers and local inhabitants. The less favourite legacies and moments of local history that are not chosen by local population are often forgotten, ignored or quite simply unsignified. The heritage landscape, throughout its long and turbulent past, has often included some unwanted aspects, which can be called 'dark heritage'. There are many ways to minimize the 'darkness' of a non-chosen legacy. The shameful heritage landscape can be consigned to oblivion, as happened with most of the communist heritages in Central Europe, often in a similar manner to how Nazi or colonial heritage is buried. Some symbols, which seem to be a part of the dark heritage for the vast majority, appear as an attractive legacy for some groups. The methods of dealing with such unwanted inheritance include assigning new, positive meanings, and substituting different, constructive or affirmative features for the negative and disgraceful ones. Many practices of reinterpretation, oblivion and de-contextualization of post-communist dark and not-so-dark inherited landscapes can be observed in every Central European country since 1989.

Chapter 3

Landscaping Socialist Cities

Landscape always shows the basic temper of the times, and, as Ruskin said, judges its character (de Botton 2007). In this sense the effect of socialist rule over Central European societies, countries and landscapes can be concluded by the socialist visual character of the cultural landscape. Ideology and urbanism have been closely entangled in even the smallest city, town and village. The landscape we now see is the result of present and past ideologies superimposed on urban tissue, and additionally modified by cultures, economies and societies. There is a strong tendency to demonstrate and perform power over people and landscapes. The tendency appears in every social system, but becomes remarkably strong in totalitarian and non-democratic regimes. Power over practically any aspect of social and economic life has to be materialized and visualized, so nobody could doubt who is in power. Every totalitarian regime of the 20th century was very much focused on transferring its powers into visible forms and meanings of cultural landscape. Mussolini, Hitler and Stalin (and many of his followers) spent lots of time and vast amounts of money landscaping the cities into a 'new, better world'.

The present order and structure of the urban landscape has been conditioned by its pre-liminal constitution, designed and conducted by communist parties and regimes. The path-dependent cultural landscapes of the former Eastern Bloc resulted both from the historical context, social and economic structure of the region, and the Marxists' vision of an ideal and politically correct landscape, infused by various efficiencies and modes of implementation. Architecture, urban planning and visual arts, together with texts and media, were seen as significant and powerful means of expression and exemplification of 'peoples' power' over the ruled masses. Centrally planned and controlled urban landscaping procedures created an advantageous environment for circulation and dissemination of chosen urban projects over all Central European dominia (Czepczyński 2006a). The totalitarian and omnipotent communist party ruled not only over the economic, social and cultural life of a specific society, but also over the visualized and aesthetic expressions of everyday existence. Communist linguistic discourses and philosophical debates over the role of means of communication, including cultural landscape, as expansive combination of form and meaning, created the significance of landscaping in its intensely philosophical context. Marxist and structuralistic dilemmas on the category of language as base or superstructure had absorbed Soviet and socialist theorists, apparatchiks and Stalin himself for decades (see Szarota 2001).

The meaningful landscape was split between the official landscape of the new cult, represented by grandiose buildings and official settings, against scenes of rather miserable everyday life, filled by constant shortages, as well as oppression and terror.

Limitation or elimination of private ownership rights gave the state or the party[1] the only real power over the landscape. The landscape was managed by number of central decisions, usually inspired and approved by political leaders. Ad hoc and off the cuff steering was the most common practice, while the permanent fear and a certain level of admiration of Soviet Union policies and expectations made the early socialist landscape management both difficult and repeatable (Domański 1997, Czepczyński 2005a). These party-states were based on fear, rejecting agreement with society. They were also characterized by deep, bureaucratic degeneration. Centralist despotism was based on a connection of mass atomization techniques with techniques of forced and multilateral organization of the masses under strict control of the state. There was a specific paradox of social consciousness: simultaneous approval and refusal of the system by many ordinary citizens (Hirszowicz 1980). Communist states oscillated between despotic, authoritarian pluralism, bureaucratic paternalism and other intrusive forms, visualized in cultural landscape features.

Socialism, landscape and power over masses

The relation between socialist ideology and landscape was deeply rooted in the basis of the communist and socialist thought. The theory of communism was principally the work of Karl Marx and Friedrich Engels, published in their Communist Manifesto of 1848. Communism is an ideology that seeks to establish a classless, stateless social organization based on common ownership of the means of production. The Manifesto specified 'the revolutionary dictatorship of the proletariat', a transitional and temporary stage between the capitalist society and the classless and stateless communist society Marx called socialism (Marx and Engels 2002). The term refers to a situation where the proletariat or working class would hold power and replace the current political system controlled by the bourgeoisie. Communism was to be the final stage in which not only class division but even the organized state – seen by Marx as an inevitable instrument of oppression – would be transcended. That distinction was soon lost, and communism began to apply to a specific party rather than a final goal. Vladimir Ilich Lenin maintained that the proletariat needed professional revolutionaries to guide it, while the other 'founding father', Joseph Stalin's version of communism was synonymous for many with totalitarianism (Communism 2007, Szarota 2001). Socialism and communism are mostly used interchangeably,[2] while during the decades of communist ideas implementation the transitional stage became for many the only and final. The distinction turned out to be merely theoretical, but very important for those who wanted to believe that they still had to suffer some more years or decades before reaching the classless utopia. The term 'Communism', especially when it is capitalized, is often used to refer to the political and economic

---

1 Party, state, government, council, committee, as well as trade union, cooperative and self-government represented the only 'right power', the main communist or workers' united party in every socialist state. The distinctions of various actors mentioned above were rather subtle, and many did not understand the differences and significances.

2 Despite theoretical differences and according to the popular use of the terms, 'communism' and 'socialism' will be used interchangeably in this book.

regimes under communist parties that claimed to embody the dictatorship of the proletariat (see Hirszowicz 1980).

The concept of socialism refers to a broad array of ideologies and movements which aim to improve society through collective and egalitarian action; and to a socio-economic system in which property and the distribution of wealth are subject to control by the community (Socialism 2007, Szarota 2001, Brzeziński 1960). This control may be direct, exercised through popular collectives such as workers' councils or indirect, exercised on behalf of the people by the state. As an economic system, socialism is often characterized by state or worker ownership of the means of production. Socialism can also be seen as a kind of extreme humanism. While the main, declared goals of socialism (human development, equal rights and equal distribution of recourses) seldom raise many disputes, the implementation of the most humanistic of the projects is always realized by faulty humans, and brings the bright ideas into the manoeuvres of the real, hard word. So one can speak about two 'socialist projects' realized in Central Europe after 1945: one deeply humanistic, referring to all the positive and positivistic aspects of social tradition, and the other one, connected with its practical implementation, based on terror, limitation of basic civil rights, and oppression. The two 'projects' are reflected in two meanings and memories of socialism, stressing one or the other parallel aspects of the system. One expression of socialism can be called 'admirative', seen as a radical humanism, while the other side is aggressive and totalitarian, focused on eliminating class enemies, ordering human existence in the smallest possible details and transforming human individuals into a 'parts of the collective' (Nawratek 2005).

Developments of the communist project

Communism or socialism started to play an important political and international role when the ideology was converted into the only authoritarian power. The process of organized, forced and institutionalized development of communism began in 1917, when the Bolshevik branch of the Russian Social Democratic Labour Party headed by Lenin succeeded in taking control of Russia after the collapse of the Provisional Government in the Russian Revolution of 1917. In 1918, the party changed its name to the Communist Party. After the success of the October Revolution in Russia, many socialist parties in other countries became communist parties, signalling varying degrees of allegiance to the new Communist Party of the Soviet Union. The highest goal of the Communist Revolution was to 'liberate' all of the working people of the world[3] from the capitalist bonds, and soon leaders of the Soviet Union and other communist parties realized that the goal could be only achieved through war (see Szarota 2001).

The Second Wold War created an opportunity to disseminate and spread communist ideas over Central Europe. Socialist or communist socio-economic system emerged in Central Europe as a consequence of the collapse of Nazi domination and seizure of the region by victorious Soviet Red Army. At the end of the War, most European countries had been devastated, and millions had been killed.

---

3 Even against their will.

Famine threatened the survivors. The Nazis and the Soviet Union had wiped out the pre-war (but not always democratic) leaderships. National communist parties moved quickly to fill the political vacuum. In almost every country liberated or occupied by the Red Army,[4] communist regimes were sent from Moscow, as in Poland and Romania, or coalition governments were formulated, with communists holding the ministries of power. The communists promised the people of Central Europe a new era of equality and economic plenty under a socialist system. Facilitated by Soviet patrons, Central European communist parties made temporary alliances with non-communists until they gained control of government power centres like the national police. The next step in enforcing the 'dictatorship of proletariat' came through uniting communist parties and was followed by the elimination of any kind of social, political and economic freedom and independence. The communist party in each country held a total monopoly of political power. This permitted no independent political parties, no meaningful elections, and no criticism of the ruling workers' party. After 1948–1949 the national communist party, under pressure from Moscow, purged any real and imagined enemies not only in economic, social, cultural and political life, but also aesthetic and artistic. By 1949, with the occupying Red Army always in the background, the communists had taken over the governments of eight Eastern European countries. The communists swiftly established so-called 'People's Democracies' in Poland, Hungary, Romania, Bulgaria, Yugoslavia, Albania and Czechoslovakia. Eastern Germany was at first a Soviet military occupation zone, but soon became the German Democratic Republic under German communist party rule (see Brzezinski 1960, Crampton and Crampton 2002, Judt 2005, Simmons 1993).

Political transformation was followed by the implementation of a socialist economic model. The government, in the name of the people, owned the factories, farms, mines and other means of production. People could no longer own their own profit-making businesses and farms, as in the capitalist system. Government economic planners decided what and how much should be produced each year, what the prices should be, and what wages should be paid to the workers. Omnipotent planners emphasized heavy industry such as steel making and coal mining. Consumer goods like automobiles, clothing and TVs became scarce and expensive. Marxists, especially those inspired by the Soviet model of economic development, advocated the creation of centrally planned economies directed by a state that owns all the means of production. Others, including communists in Yugoslavia in the 1960s, Hungary after the 1970s and Poland in the 1980s have proposed various forms of market socialism, attempting to reconcile cooperative or state ownership of the means of production with market forces, which guide production and exchange in place of central planners (see Lovenduski and Woodall 1987, Judt 2005, Simmons 1993). During the second half of the 20th century, many other countries adopted

---

4 The only exceptions include Austria and the Danish island of Bornholm. Although the Soviet troops occupied Bornholm, Vienna and eastern part of the country, Austria managed to escape the fate of the would-be Eastern Bloc. At the same time, Yugoslavia and Albania had not been captured by Soviet troops, but their leaders chose communism; though without pressure from the Red Army, they were not forced to fully follow the Soviet social and economic model.

a pro-communist government, often supported by Soviet help or various kind of assistance. At some point Mongolia, China, Cuba, North Korea, Vietnam, Laos, Angola and Mozambique were 'building socialism' more or less enthusiastically. By the early 1980s almost one-third of the world's population lived in Communist states.

Despite the variations of communist implementations, wherever the communist party took over power, they were never able to risk democratic procedure to verify the operation and results. As the experiences of Central Europe, Soviet Union and other communist countries have taught that having communists in power always eventually leads to totalitarianism, oppression, limitation of freedom and authoritarian regimes built on false imaginations and military forces. The autocratic system was facilitated by usually very efficient propaganda and indoctrination. Full control over any media, including landscape, kept societies under low aspirations and limited dreams.

Landscape implementations of communist ideas

The questions of spatial organization, urban planning and eventually landscape were often discussed by many representatives of the communist elites, including the 'founding fathers' (Engels, Marx, Lenin and Stalin). Most of them were to criticize the city as bourgeois and the source of most contemporary social problems. Engels, on one hand, found cities to be the site of the main concentrations of the proletariat and breeders of revolution, but on the other hand he also suggested de-urbanization and the gradual liquidation of large cities (Goldzamt and Szwindowski 1987). Landscape was an important tool for empowering the communist rulers. Forms, functions and meanings of urban landscape appeared to be the perfect medium to communicate the relationship of powers, as well as the new aesthetics and styles. Landscape was also monumental, vast and relatively easily understood by the masses. State socialism was characterized by a general ideologization of practically every aspect of social life, and there was no exception of urban and cultural landscape.

One of the main fundaments of socialist architecture was the will to change society by or through, architecture, design and cultural landscape. The main goal of the Soviet, and then all socialist, architecture and urban design was the pursuit of the fullest possible human development, as the highest value of socialist society (Szyszkina 1981). There was a strong structuralist belief that social and living conditions create the individual, his or her personality and value system, lifestyle and construct the entire personality. In this context, the creation of a visual environment became an integral and important part of the creation of a 'new man'. There was no place for 'bourgeois freedom', but a fight for socialist man living in socialist houses in socialist cities (Nawratek 2005). Since working people were now the ruling class, then why not build vast hive-palaces for them, and transform peasants into aristocrats? The promise of the state communism found architectural expression throughout the 1950s, from the Moscow underground to the Warsaw Palace of Culture and Science (see Wagenaar 2004). The conceit is obviously a totalitarian version of 'let them eat cake', or 'bread and circuses', aimed at satisfying the crowds and admiration of the 'new Caesars' (Jencks 2005, 25–26).

Socialist architecture and landscapes had been created in relation to both aspects of socialism: the humanist ideology of helping the poor and the Marxist oppression machine. There have been constructed many landscape features aimed to correspond to the social aspect of the communist ideology, like mass housing, cultural and sport facilities, schools, technical infrastructure and many others. Most of the newly constructed features simultaneously carried ideological aspect of the much-needed function, focused on realization of the 'humanistic project'. The other, 'aggressive' and dominative face of communism was also visualized in urban landscape, usually by grand and overwhelming constitutions, expressing the structure of power, allies and history. Communist landscape reflected all major concepts of the system of thoughts, and followed all the periods of 'mistakes and perversions'. Despite of the efforts of the decision-makers and planners, information was often not clearly transferred and left the field for re-interpretations and mis-interpretations. Generally poor codes and a reduced capacity of the codes, together with usually low quality of the realizations, made the implemented ideas ironic icons (see Simmons 1993, Judt 2005).

Many of the public buildings of the totalitarian regimes, especially of the 1950s, can be best approached as propaganda, not simply by virtue of the insignia, heroic narratives and inscriptions inside and out, but also of the expressive form of the architecture. At the heart of the socialist dictatorial regimes was a very deeply coded need to be familiar as well as autocratic, to be 'of the people' but not of a specific region or class. Some of the conditions of the 'regime architecture' were purely technological, while others were more socio-technical. Communication with vast, agitated crowds required a language that could be broken into segments, articulated in bold, abstract bursts, placing great stress on single words (Benton 1995). The language also had to be familiar to the usually uneducated masses, so many well-known sacral metaphors appeared in cultural landscape of the Central Europe to emphasize the significance of the new powers in popularly understand spatial language.

Sacrum, myths and cult

Socialism, as ideological system, was to some extent based on various myths, connected with rites (Łukasiewicz 1996). Socialist rites required objects of celebrations and particular spaces of celebrations. Both were created in socialist cities: the socialist 'gods' had been produced, together with a pantheon of socialist heroes, celebrated according to ritualized cult (see Unfried 1996, Satjukow and Gries 2002). The socialist pantheon can be divided into two categories: allegoric abstract concepts and idolized people (Rembowska 1998). Revolution was considered the prime 'god' and divided the universe into bad (before the Revolution) and good (which came after it). Many monuments and sanctuaries had been erected to worship Revolution, often in a dominant position in urban structure, located in the main squares or exposed on the border of green areas. The Party[5] was the other abstract

---

5 Even though there were usually a few, often three, official political parties in every communist country, only one, the leading united, communist or workers' Party, was really important.

'god', present in socialist landscapes in 'houses of the Party' as its local headquarters had been ephemerally called. The other category of celebrated people included the 'founding fathers' of the communist system – Marx, Engels, Lenin and Stalin.[6] Their cult was hierarchic, at times changing, and supplemented by dozens of other, lesser heroes, including military leaders, local martyrs and communists. Many schools, streets, and factories were named after them, while their official portraits decorated streets and offices.

Socialist myth made a mythic space and time (Łukasiewicz 1996). Socialist mythic space had a dichotomic structure: separating the good space from the bad one, the world of progress from the world of backward. The 'good space' was associated with building a new, better world, work for advancement and the Party. It was the space of grand socialist designs and constructions, industry, infrastructure and housing. The 'good' spaces were in opposition to the negative, pre-socialist ones. Those spaces were testimonies of the past, fallen systems and were adapted to new functions, usually as places of production, power, education and culture. Often the old 'witness landscapes' were demolished or left to slowly decay. Many 19th century capitalist urban landscapes, imperial, royalist or aristocratic palaces and manors, churches and monasteries disappeared from the Central European landscapes. The mystic spaces were best seen during celebrations of many, newly established feasts. The celebration required usually large, open spaces. Grand squares in almost every socialist city centre were not only signs of lack of land rent, but also played important ideological functions. Marches, manifestations, meetings, speeches and parades were a crucial part of socialist ritual and ideological celebrations. Those celebrations created new sacrum in socialist cities, where the cult to the abstract or personal 'gods' was ritualized (Rembowska 1998, Satjukow and Gries 2002).

Communist system of representations and rituals can be compared, to some extent, to a religious cult. 'Gods' and 'saints', as well as manifestations, were celebrated and commemorated at places and spaces of *sacrum* and *profanum*. Special roles were played by exclusive communist *charkas* or a nexus of metaphysical and transcendent energy. The holiest of the holy in all communist universe was of course Lenin's mausoleum in Red Square in Moscow. The local 'middles' included the Dimitroff mausoleum in Sofia, the copy of the balcony from which Karl Liebknecht declared the socialist republic in Berlin in 1918 (Figure 3.1), and the Palace of Culture and Science in Warsaw. There were many legends and myths about the secrets of the ruling class and their confidential and mysterious landscape features and safe, underground world. Classified passes, bunkers and underground towns were developed not only in popular gossip and imaginations, but to some extent actually existed under the capital cities.

---

6 Stalin was among 'the highest gods' of the communist pantheon only until the late 1950s.

**Figure 3.1 Communist sacrum: 'historical' Karl Liebknecht balcony of the
        Berlin Castle, 2005**

There was another myth of the socialist city, which grew from the lack of free media, critical discussion and comparison, and because of travel limitations. Many of us believed that we actually lived in the best possible system. The myths of perfect spatial and social organization, balance and sustainability, and union in mass had a very strong hold over and were popular in socialist societies, despite often being negatively verified by common sense, they were at the same time believed and appreciated.

Planning, designing and realizing

Socialist economic planning began soon after the war in most Central European countries with the introduction of reconstruction plans, usually of two or three years' duration. Next came the five-year plans. The plans shifted investment and labour into industry, especially in the heavy industry sector to the cost of agriculture and consumer goods. There were massive planned increases in production, in industrial, construction and agriculture (Crampron and Crampton 2002). The centrally constructed plans soon became the main economic tool, aimed at replacing the supply–demand relationship typical of market economies. National, regional and sectorial plans developed into acts of law and became the main mechanism of market regulation, while the system was converted into a centrally planned economy and society. Since technically everything had to be planned, special roles were seceded into the planners, designers and architects of the new economy and reality (Judt 2005).

Architecture was an extremely important weapon is the hands and heads of the creators of the new social order. It was intended to help to form a socialist theme – enlighten the citizens' consciousness and change their style of life. During this great work, a crucial role was left to the architect, who wasn't perceived as merely an engineer creating streets and edifices, but an 'engineer of the human soul'. Under socialism, architecture and cultural landscaping became an important social mission, while the architects and planners were a notable professional group, which often, with modernist megalomania, thought of themselves as the 'Demiurges of the space'. Architecture and planning were deeply involved into the state socialist system and the ideology of reconstruction of society towards the elimination of social differences and contradictions. They became 'missionaries' or 'priests' of that ideology (Nawratek 2005). Socialist architects and spatial planners believed that architecture was the privileged art, which can change life and living conditions. The concept of *l'architecture parlante* was especially evocative and vivid within the high hierarchies of communist parties. As declared at the communist architects' meeting held in the building of the Central Committee of Polish United Workers' Party in 1949, 'architecture must become element of socialist education of the masses' (Goldzamt and Szwindkowski 1987). Architects became the privileged designers and coders of power in both the urban and rural landscape.

Many 20th century dictators had architectural/artistic aspirations and definitely appreciated the significance of landscape. One of Stalin's favourite titles was 'the grand constructor', and he loved to spend time designing or imagining views of new cities.[7] Many local leaders followed the great example from Moscow. The question of architecture and meaningful landscape was among the most important issues, especially for Nicolae Ceausescu in Romania or Bolesław Bierut, the president of Poland, 1947–1952. Polish leaders of the communist party and the president in the early 1950s paid lots of attention in designing Warsaw. Officially, Bierut made himself the only author of the grand Warsaw 1952 Master Plan. The Political Bureau (*Politbiuro*[8]) of the Central Committee of the Polish United Workers' Party, which was the most important assemblage in the country, took many of the general and detailed cultural landscape decisions. On 24 March 1951 the *Politbiuro* decided that the newly designed monument of Felix Dzierżyński should be placed 'on the main axis of middle risalite of the main pavilion's western wall, 20 m. east of the wall' (Dudek 2005, 215). The whole meeting on 3 January 1952 was focused only on 'architecture and urbanism of the Muranów district of Warsaw, particularly Southern side of Leszno Street, between Dzierżyński Sq. and Żelazna St' (Dudek 2005, 216). Three weeks later the discussion was focused on the shape of the street lamps on the Constitution Sq. After rather long deliberations, a project was approved of '8 lamps candelabra, in the form of spire obelisk' (Figure 3.2) (Dudek 2005, 216).

---

7 Similar to Adolf Hitler's ambitions of remodelling Berlin into *Germania*.
8 *Politburo* is the Russian abbreviation of 'Political Bureau', the most important body of central committee of every communist party in the region.

The main focus was on the form of architecture, together with the function, but altogether only in the sense of 'socialist meanings' and significance. According to Bierut's personal comment, the Nowa Huta town hall tower, 'as a vertical dominant, should be particularly carefully elaborated'[9] (Dudek 2005, 216). Bierut's successor, Władysław Gomułka, evinced no interest in questions of architecture or design, but concentrated on economic issues. After the 1960s architecture was to a great extent left to the architects, dominated by the notion of Corbusierian International Style, as in the rest of the world.

**Figure 3.2 Importance of the details: elaborate candelabras in
Constitution Sq., Warsaw, 2007**

9 Probably the elaboration went too far or/and in the wrong direction, since the Nowa Huta town hall was never constructed; there is a definitely not 'carefully elaborated' district green in the place reserved for the town hall.

Political decisions had been put into operation by planning offices. Architects were often seen as collaborators with the regimes and at least partly guilty for the miserable and dysfunctional urban landscape. Modernist design was considered treason in the late 1940s and early 1950s. Architects designing against the Party's wishes might be jailed for years, sent to Siberia, tortured, or sent to labour camps. Others, either out of fear or for benefits and fame, followed any instructions given by the Party, sometimes even to a terrifying extent. But many architects and planners deeply believed in their mission of realization and implementation of a 'new, better world'. For many of them, new regimes created new opportunities to realize their social and modernist ambitions and visions. Since the late 1950s, after the dark ages of socialist realist historical styles, heavily inspired and/or forced by Moscow, architects enjoyed a relatively high level of creative freedom, sometimes to the level seldom seen in a market economy. Any priority project, approved by the Party, had hardly any limitations: costs, land and materials – everything had to be, and actually was, organized. Since a building, project or landscape represented the communist power and party, any obstacle or difficulty was translated as acting against the party and, sometimes, against socialism, and was treated as a serious crime. Many architects used the new opportunities to realize enormous modernist landscape projects of blocks of flats, together with public and engineering buildings.

Ambitious and grandiose plans were rarely fully implemented. Constant shortages, bad management and sudden shifts of priorities together with regularly limited assets made most grand designs half-done. Many housing estates lacked non-essential services: very few of the planned culture centres, administration buildings or sport facilities were completed. One of the characteristic and distinctive elements of a socialist city is urban wasteland, appearing in almost every city in various forms. No land rent or real land owners and lack of market pressure over 40 years created huge open spaces, unnecessarily wide roads, and immense squares. The openness and deurbanized spaces can be seen not only in peripheral districts, but also in city centres. Uncompleted projects became typical 'trade marks' of socialist landscapes. Hundreds of half-built and/or derelict roads, bridges and buildings can still be seen in villages and towns all around the region. Even the priority grand projects, like Poruba, an eastern district of Ostrava, now in Czech Republic, were never completed. The grand crescent arrangement[10] seems to be uncompleted and surrounded by unkempt urban green wasteland. Similarly the main square of Nowa Huta in Kraków, Poland, has lacked one of the façades, since the location was reserved for municipal authority buildings in the early 1950s.

Socialist hierarchy of places

Making socialist landscape significant and controlling that significance was one of the important tasks of the new communist regimes. Ideological features of cultural landscape can be implemented on many different levels, incarnations and manifestations. There was a common, but not officially recognized hierarchy of places, buildings and features, which constructed the official landscape of power.

---

10 Which seems to be inspired by Bath Royal Crescent.

Different landscape and socio-techniques were employed to facilitate desired meanings and enhance the power of the Party. All three classifying codes of meanings (see Eco 1985), typological, spatial and syntactic, were used by communist decision-makers and architects.

Typologies have been connected with function, and the most important function of communist cities were in their centres of political power, always located in the central committee of the Communist Party's headquarters. In smaller towns, the committees were sometimes located in the same building as municipalities, but in larger cities the Party had a separate, usually newly constructed building. Communist Party quarters, at national, regional or very local level, were clearly the focal point of powers in landscapes. Often structures of insignificant form concentrated extensive control. In every country of Central Europe, Central Committees of the ruling parties had been located in representational buildings in most prestigious locations. Most of them are classicistic constructions from the 1950s[11] and often located on the river, as in Berlin, Prague and Budapest (Figure 3.3). Rather modest, but stable and substantial constructions looked as solid as the system itself.

Figure 3.3 Centre of communist powers: former communist headquarters, Prague, 2007

---

11 Except for Berlin. Pragmatic Germans were the only ones to adapt an older building, the former 1940s *Reichbank*.

Syntactic codes had been implied by many different modes, including celebration of heroes and 'sacred' events. Importance was enforced and enhanced by significant names and episodes answering the growing demand for new names of factories, streets, and new towns. In search of 'proper' names and codes, communists incorporated almost all leftist, social democratic, socialist and workers' rights traditions, heroes and heritages. They assimilated all the 19th and early 20th century protagonists and activists as theirs. Their graves were turned into shrines. The biggest and most famous was 'The Monument of the Socialists' at the Central Cemetery Berlin-Friedrichsfelde, where all of the most prominent and German communists' bodies were gathered.[12] Hundreds of memorials and monuments were raised, not only to the great leaders like Lenin or Stalin, but also to numerous national heroes and idols. Sometimes the chosen protagonist was hardly a communist, but become a common icon, like the 19th century Polish socialist Ludwik Waryński, who was made patron of hundreds of streets and a large mechanical factory, and was an important component of the communist propaganda scheme. Dictatorship of proletariat implied also dictatorship of designation and denomination of places. All of the names connected with the past system had to be erased and replaced by new patrons. There was a hierarchy in place naming: the biggest 'saints' could be patrons only of the main squares, streets or largest factories. So Lenin, Stalin, Revolution, Liberation, and Victory streets were unmistakably central and important in many towns and cities of Central Europe. Additional coding was connected with monuments of the liberating Red Army. Since Soviet troops captured all of Central Europe, thousands of war graves and memorials had been erected to commemorate the dead soldiers, as well as to strengthen the 'eternal friendship' between the Soviet nation and the nations of the 'brotherhood democratic republics'.

Significant socialist spaces had to be constructed for public manifestations supporting the regimes. Regularly organized demonstrations backing the system, meetings, gatherings and marches created demand for spacious streets and squares, which were very important modes or axes of spatial organization of socialist cities, and were particularly visible in large or new cities. These square were only a container for the anonymous crowd, ready to cheer and greet its leaders. Grand 'magistrale' or main alleys had been designed or chosen to perform in major 'sacral' celebrations, including 1 May and many others.[13] The concept of a large public square used to serve as a designed agora-like space. On those vast squares and avenues masses could and often had to worship and/or admire their rulers. Since open spaces were not really used for democratic practices, but only to enhance the communist power, those places became *anti-agoras* of socialist cities, where the people could only manifest their support to their representatives (see Smith 1996). Moscow's Red Square, Berlin's Stalin Avenue, or Warsaw's Square of Defilades had

---

12 Including Rosa Luxemburg, Karl Liebknecht, Otto Grothewohl, Wilhelm Pieck, Ernst Thälmann and Walter Ulbricht.

13 The manifestation and the march was always planned and organized, and happened in even the smallest town and city of the Bloc. Of course participation in the event was only formally voluntary, but absence could have financial and disciplinary consequences for employees, school children or their parents.

been typical *anti-agoras*, where one as a part of the mass could/should reveal his or her collaboration with the system (Nawratek 2005). The spaces of manifestation were usually located just off the main Party buildings. The *anti-agoras* created best possible mis-en-scene for 1 May, 9 May, Revolution Day, National Day[14] or any, often obligatory, gathering. Sometimes, contrary to the intentions of the communist planners, spacious *anti-agoras* became real *agoras*, where people manifested their disappointment against ruling regimes, as it happened on Stalin Allee on the 17 June 1953 during the Berlin uprising, or on mains streets of Poznań throughout the June 1956 riots.

Another important function and feature of socialist landscape was culture. On one hand we had the official, controlled and Party facilitated 'state culture'. The official culture was extremely important for the regimes, as a major ideological and propaganda tool. Actors, writers, musicians, painters and architects belonged to the very privileged class, and had much more freedom and possibilities then most of us. Control and edification through culture and its institutions had been an imperative part of socialist place and time making. The main representation of official 'culture cult' were impressive, sizable, significant culture centres, known as 'palaces' or 'houses' of culture. These important features were built in almost every major socialist city. The grandest include, in the early 1950s, Warsaw's Palace of Culture and Science, and in the late 1970s, palaces of culture in Prague[15] and Sofia (Figure 3.4). The buildings, often in quite sophisticated forms, hosted wide varieties of cultural establishments, including conference and music halls, museums, libraries, together with many various cultural organizations, and sometimes a theatre. The buildings were also representational, in double meanings. At first they represented the attention to and importance of culture by communist parties. The second type of representation was connected with form and urban setting: often centrally located or well exposed, the 'palace of culture' was location of almost religious 'culture cult'. The other sort of culture comprised of less legal cultures, partially independent and somehow contesting the regime, in direct or indirect way. The selection of theatres, galleries, churches offered some alternative cultural messages, usually deeply coded between texts, paintings or music (see also section on Non-socialist features of socialist cities, below).

Sometimes landscape features were built mainly for propaganda reasons to show off or hide unwanted structures, like churches or monuments. The most famous example of using buildings for political reason was the housing development on Leizpiger Street in central Berlin. In 1966 Axel Springer, the owner of the *Bild*, had his press building constructed in West Berlin right by the Wall. The golden brass façade of the 68-metre high tower was clearly visible from most of the centre of East Berlin. The roof of the media building had a neon display visible to East Germans with daily news from the free press. The GDR government answered with the

---

14 The communist era main national celebration was usually related to the day the first communist government was established or legitimized around 1945. For example, Poles were forced to celebrate 22 July 1944, when the communist provisional government was brought from Moscow to Lublin, eastern Poland.

15 Now it is a half-used congress centre in district of Vyšehrad.

construction of four 25-storey apartment buildings near Leipziger Street, intended only to block the view of the Springer news (Klopeck 2006).

Figure 3.4 Culture in landscape: House of Culture, Sofia, 2005

Industrialization and the dictatorship of production

All countries of Central Europe[16] entered the socialist period in the late 1940s as generally rural and non-industrial societies and economies (Judt 2005). Industrialization and modernization became of the main goals of the new governments. Post-war regional economies were mostly wrecked, with many cities destroyed, so demand for industrial, and especially heavy industrial products was enormous. Dictatorship of production became a credo for every socialist economy (Domański 1997). One of the main characteristics of the communist system was massive and rapid urbanization and industrialization. The planned economy prioritized industrialization, and the intention was not purely economic. Grand industrial plants were not only to produce much-needed steel for more construction and tanks. One of the main goals of forced industrialization was to produce new, class-conscious workers, as a final product of socialist social revolution, and was also 'to create an industrial infrastructure which would produce the proletariat for whom Communist

---

16 Except for Bohemia in Czechoslovakia, Thuringia with Saxony in East Germany and Silesia in Poland.

Party rule was designed and which, according to Marxist theory, would inevitably inherit the earth' (Crampton and Crampton 2002, 159). Industry, especially heavy steel works, power stations and chemical plants became an economic and social priority, while consumption production would wait until better times.[17] The plans set out to expand the energy producing and metallurgical sectors of the economy. Steel was accorded a special significance and propaganda efforts were devoted to steel production, especially in newly founded plants, like Eisenhüttenkombinat Ost in the eastern part of East Germany.

Box 3.1 Eisenhüttenkombinat Ost

The third congress of the Socialist Unity Party of Germany, 20–24 July 1950, decided to erect a steel mill, the Eisenhüttenkombinat Ost, and an adjacent residential city. Construction began on 8 August 1950. The first blast furnace was put into operation one year later. The residential city was given the name Stalinstadt in honour of Joseph Stalin in 1953. The city is also ironically known as *Schrottgorod*.[18] As a consequence of de-Stalinization the city name was changed to Eisenhüttenstadt on November 13 1961. Eisenhüttenstadt was advertised as the 'first socialist city on German soil'. Like other new socialist cities, such as Nowa Huta in Poland, it followed the example of Magnitogorsk in the Soviet Union and was built alongside a new state combine. The city plan was designed by the architect and planner Kurt W. Leucht. On 1 January 1969, the Eisenhüttenkombinat Ost together with other steel manufacturing enterprises was consolidated into the state-run VEB Bandstahlkombinat 'Hermann Matern', named after prominent East German politician. After German reunification the VEB Bandstahlkombinat 'Hermann Matern' was renamed 'EKO Stahl AG' and prepared for privatization. Due to increased competition from West German steel makers and the collapse of markets in Eastern Europe, the EKO Stahl AG had to lay off workers and close several blast furnaces. In 1995 the steel mill was privatized and sold to the Belgian steel maker Cockerill-Sambre, now part of Arcelor Mittal (Eisenhuetenstadt 2007).

Similar ideologically enhanced steel works had been constructed or enlarged all around the region. In Poland these were Lenin Plant in Nowa Huta (Figure 3.5) and Beirut Mill in Częstochowa, together with Gottwald Steel Mills in Kunčice in Czechoslovakia, Lenin Plant in Pernik in Bulgaria (Crampton and Crampton 2002). The plants were not always economically feasible, while five year plans enforced concentration on the quantity rather then quality of production.

---

17   Which, at least in Poland, never came during the communist period.

18   It is humorous a compound of the German word *Schrott* for scrap metal and the Russian word *gorod* for city.

Figure 3.5 Industrial new town: administration building to Nowa Huta Steelworks, 2006

Industrialization was also used as a tool and means for reinterpreting the city. Rapid development of heavy industry in the old capital of Poland, Kraków, was to some extent a communist answer to the old, university, bourgeois and clerical image of the local society. Industry was a tool to reshape the city in economic, architectural and, most importantly, social aspects.[19] Thousands of uneducated workers, usually from rural areas of central and eastern Poland migrated to the new, specially designed model quarter of Nowa Huta. The newly urbanized working class was expected to change the old, unwanted social structure, while the grand Lenin Steelworks were supposed to become new symbol of the city. Powerful plants carried various functions, not only manufacturing. The paternalistic model of industrial duties had been fully implemented in practically every larger factory. Many facilitated free-time activities of their employees. Some of them, rather ironically, became cradles of a real 'new working class', highly ungrateful to the communist regimes, as happened in Gdańsk (Lenin) Shipyard (Figure 3.6).

19 It is said that a referendum held by the authorities was soundly defeated by the people of Kraków – a major cause of embarrassment for the government. To 'correct the class imbalance', the authorities commenced building a satellite industrial town to attract people from lower socioeconomic backgrounds to the region, such as peasants and the working class.

Figure 3.6 Iconic industry: Gdańsk (Lenin) Shipyard, 2005

Five-year plans had been overwhelmingly concentrated on production and had little or no scope for environmental protection. In consequence, rapid and quantity-focused industrialization caused degradation of the environment, often exceeding that of Western Europe. The prime cause of this despoliation was the regimes' obsession with plan-fulfilment, no matter what the cost. As a result, no filters were fitted into factory chimneys, there were no waste-water processing plans and little if any restraint was imposed on the disposal of even toxic waste (Crampton and Crampton 2002). Industrial landscape became dirty and polluted, often much worse than in the late 19th century. The most affected regions of Central Europe were those most industrialized, including Saxony and Thuringia in Germany, northern Bohemia in Czechoslovakia and Silesia in southern Poland.

Box 3.2 Lenin Shipyard, Gdańsk

The Gdańsk Shipyard was opened in spring 1945, based on the nationalized former German Schichau and Klawitter shipyards. In 1967 the shipyard was awarded and honoured by the name of V.I. Lenin, undoubtedly a grand privilege and mark of distinction. The official reason to choose such a patron was a commonly known fact, that Lenin 'achieved a success, far more important than any previous triumph of human genius'[20] (Fortuna and Tusk 1999, 172). The shipyard was the largest shipbuilding company in the Eastern Bloc, employing at its height about 16,000 workers. The plant was the pride of the regime and often represented the region outside. Practically every important visitor to Gdansk had to pay a visit to the shipyard, including all high-ranking officials. In 1975 the shipyard was awarded the prestigious First Class Workers' Banner Honour, and its big copy is still hanging on the wall of the largest shipyard hall. The shipyard, as with any grand entity, developed an extended system of social institutions, including its own internal transportation, telecommunication, health care, sport clubs, education, recreation and housing. The shipyard, as a massive conglomerate of privileged labour, created an opportunity to develop an independent, active and self-conscious social class. Unfortunately for the communist party, the self-conscious class turned against the 'feeding hand of Mother-Party'. Beginning in about 1970 the shipyard became a cradle of independent and illegal trade unions and human right activities. Those activities resulted in massive strikes and riots in August of 1980, led by Lech Wałęsa. The shipyard was the leading force of social and political transformation of the country and later the whole socialist bloc. The free market shock of summer 1989 saw the Shipyard unprepared for new challenges. The old supply and demand chains with the Soviet Union were suddenly broken and the factory was left on its own. The political heritage of the plant did not help a lot, while the political support of former worker, then president of Poland, Lech Wałęsa brought some hope, but not the much needed investment. Potential German, Norwegian and American investors were nervous of omnipotent trade unions and high social expectations. The final act of bankruptcy was declared in 1998. The shipyard was overtaken by Gdynia Shipyard and developer Evip Progress Company, later Synergia 99. In 2007 the shipyard employs approximately 2,000 people and has been bought by a Ukrainian investor, while part of the former shipyard is being prepared for water-front developments, a few buildings of the old complex house the Solidarity Museum, and few more are leased to local artists (Czepczyński 2005b).

New urbanism and new towns

Massive industrialization and urbanization in all Central Europe created massive housing demand, accelerated by mass destruction during the Second World War, especially in Poland and Germany. New towns and new districts had been built, not only to accommodate the new and growing population, but also to represent new powers and new social relations. Hermann Henselmann, chief designer of the Stalin Avenue complex in East Berlin said the in late 1940s that in East Germany 'grew especially strong demand for everyday social contacts (...). Now urban and architectural

---

20 Although, unfortunately, nobody could precisely tell what kind of masterpiece of minds it was.

environment will favour social life and cultural interpersonal communication (…). Clear, suggestive forms were aimed to create urban atmosphere, supporting person's identification with society, by identification with the city and new, democratic state' (Goldzamt and Szwindkowski 1987, 149). Building a new town was often compared to creation of the universe. The process began in the late 1940s and was focused on creation of 'brand new world' from the darkness of pre-communist/capitalist chaos. The ordered, communist cosmos was in opposition to the chaos of bourgeois landscapes. The mighty Party, through their planners, made the new, better world, usually built in the form of grand deigns. At the same time the working class masses entered the city centres and upper class residential districts, as well as universities, offices and cultural services, not only as users, but also as facilitators and new decision-makers. New and often barely educated landscape lords directly or indirectly influenced new urban designs, while 'new tastes' were enthused by unsophisticated historical and rich styles of classical or baroque old churches and palaces.

The socialist city or *socgorod* was not designed to live a good life, but rather to sustain industrial development. Socialist powers were coded, or rather hidden behind statements, norms and documents, but at the same time nobody doubted that the power was there, even if it could not be directly seen (see Smith 1996). Most of the new towns had been constructed to accommodate the 'working force' of the grand industrial plants, preferably steelworks, like in the Hungarian town of Dunaújváros.

Box 3.3 Dunaújváros

After the Second World War the new, Communist government started a major industrialization programme in support of its rearmament efforts. In 1949 Dunaújváros was chosen as site of the largest iron and steel works. Originally they were to be built close to Mohács, but Hungarian-Yugoslavian relations worsened, and this new site was chosen, farther away from the Yugoslav border. Initially the town was designed for 25,000 residents. The construction of the town began on 2 May 1950 near Dunapentele. Within one year more than 1,000 housing units were built and the factory complex was under construction. The town took the name of Stalin officially on 4 April 1952; its name was Sztálinváros, or 'Stalin Town', parallel to Stalingrad in the USSR. The metal works were opened by 1954. The town had a population of 27,772 at this time; 85 per cent of them lived in comfortable apartments, while about 4,200 people still lived in barracks which originally provided homes for the construction workers. In the middle of the 1950s, public transport was organized, and buses carried 24,000 passengers each day. During the 1950s many cultural and sports facilities were built, and the Endre Ságvári Primary School was the largest school in Central Europe in the 1960s. In 1956, the construction was hindered by an earthquake and a flood, and in October by the 1956 Hungarian Revolution. During the revolution the town used its historical name Dunapentele again, and after fell under martial law. After the Revolution the town was still the 'trademark city' of socialism in Hungary, and was presented as such to foreign visitors. Among the visitors were Yuri Gagarin and the Indonesian president Sukarno. The city also provided scenery for popular movies. On 26 November 1961 the town's name was changed to Dunaújváros, new city on the Danube. The DUNAFERR factory complex is still is a determinant enterprise in the Hungarian steel industry, and a major employer in the area. Today 'New Danube City' is home to many new infrastructures and the new South Korean Hankook tire factory (Dunauvaros 2007).

Socialist landscaping was radical break with the old shape of the city, as if trying to eliminate any path dependency and forgetting the old developments. The socialist city, especially in its initial, Stalinist form, was to become a *Gesamtkunstwerk*: a total work of art, based on a 'harmonious' interaction between urban composition, architecture and other arts, or, what we can call now, holistic aspect of cultural landscape (Groys 1988). There is an echo in socialist urban landscapes of the illusory canvas villages built for Catherine II by Count Potiomkin. But in this case the illusion of the façade was compounded by the use of the entire on-street building itself to create the illusion of an imaginary city, concealing the prosaic substance that survived behind (Cohen 1995). The monumental landscapes include many grotesque examples of 'speaking architecture'.[21] The search for an 'assimilable' past-prescribed by socialist realism as the only way to make the city legible for the 'masses' – gave rise to a wide range of sometimes surprising architectural inventions, like incorporating pseudo-baroque or renaissance décor. The new socialist cityscapes were transformed by the vertical buildings, as modern counterparts of the towers and spires of ancient churches and castles (see Figure 3.7) (Cohen 1995).

Figure 3.7 New town for new man: Poruba, Ostrava, 2007

---

21 Probably the most famous examples of Stalinist *architecture parlante* is the Red Army Theatre in Moscow, whose overall plan and cross-section of its pillars are in the shape of a five-pointed star (Cohen 1995). The other illustration is Ploshchad Svobody or Freedom Square in Ukrainian city of Kharkiv, designed in the shape of a sickle and hammer.

As an answer to growing housing demand, many countries introduced the special and restrictive housing policy. Re-evaluation of aesthetic canons made a convenient climate for the new International Style in architecture. Additional requirements came from the rapid urbanization and industrialization in the 1960s and 1970s. Constructivism in architecture and urban landscaping was followed by the concept of 'constructing' new, better, socialist man and society. Landscape was only a tool, one of many, together with education, media, industrialization, cultural policy, collectivism and many others developed to 'construct' new social conditions of circumstances, which would create new personality and new patterns of political and social behaviour (see Ortega y Gasset 1994).

There were numerous new towns and large urban districts constructed or largely developed in communist times in soc-realistic and modernist manners. The most famous examples of socialist town planning include Eisenhüttenstadt and Schwedt in East Germany, Dimitrovgrad in Bulgaria, Dunaújváros and later Leninváros in Hungary, Havířov and Most in Czech Republic, together with many Polish towns, like Kraśnik Fabryczny, Płock, Puławy, Bełchatów, Tychy, and probably best known and largest, Nowa Huta.

Box 3.4 Nowa Huta, Kraków

The biggest new town in Central Europe was started in 1949 as a separate town near Kraków. It was planned as a huge centre of heavy industry. The town was to become an ideal town for the communist propaganda and populated mostly by industrial workers. In 1951 it was joined with Kraków as its new district and the following year tramway communication was started. On 22 July 1954 the Lenin Steelworks was opened and in less than 20 years the factory became the biggest steel mill in Poland, employing at its height about 40,000 workers. In the 1960s the city grew rapidly. The architectural concept of the new town was prepared by Tadeusz Ptaszycki and the team in 1949 and was based on renaissance, baroque and garden towns' concepts: streets spread radially from the Central Square, and buildings are located along a main axis. The monumental architecture of the Central Square (Plac Centralny) was surrounded by huge blocks of flats. After 1965 modernist blocks of flats and infrastructure facilities had been built. Despite of many obstructions, supported by street fights and rebels, a Catholic church was constructed in 1967–1977 during a 10-year-long period of negotiations with communist apparatchiks. The Lord's Ark temple resembles Le Corbusier Ronchamp chapel and became a new symbol of hope, faith and victory over socialist apparatus. During the Solidarity Revolution of the 1980s Nowa Huta became one of the major anti-communist resistance centres. After 1989 the steelworks lost its significance, and the streets formerly named after Lenin and the Cuban Revolution were renamed to honour Pope John Paul II and the Polish exile leader Władysław Anders. In 2004 Plac Centralny, Nowa Huta's central square, which was once home to a giant statue of Lenin, was renamed Ronald Reagan Central Square in honour of the former US president. Nowa Huta's population is about 220,000, and has become quite a tourist attraction of soc-realist urban design (based on Nowa Huta 2007).

Projects in socialist new towns were hardly ever completed. Public ownership of the land generated no market pressure, so many of the new towns are characterised by vast, open spaces, even in central parts of urban establishments. Additionally, most often those towns lack the most important part: a multifunctional town centre. The centre might have been important for town life, but was much less important for the steel production. The towns were generally monofunctional, based only on heavy industry, mining or chemical plants. Practically all of the housing stock remained city or factory owned so instead of modern, multifunctional urban structures, typical factory towns were created instead. But probably the biggest failure of the socialist new town project was connected with fiasco of 'making new man'. Workers and their families were as socialist as anywhere, and often new towns were concentrations of crime. Making 'working class man' projects often were very successful, although in contrary to the regime's expectations. What was even worse for the ideological planners, sometimes like in Nowa Huta, those 'new men' of 'new towns' initiated anti-communist resistance and strikes. The 'new workers' were often class-conscious and revolutionary, fighting for the freedom of trade unions and against low salaries, but this time against communist authorities.

Historical policy of landscape

Building absolutely new systems, which rejected any previous achievements and developments, somehow implied and preferred 'green field development', while the old, used 'brown fields' would be eliminated or marginalized. Nationalization and elimination of unwanted features, including royal, bourgeois, religious and ethnic minorities were deeply coded into the 1950s cultural landscapes. The historical policy of landscape can be characterized by Lenin, who emphasized the role of selection: 'not only pure negation, hesitation, doubts are characteristic and important for dialectics, but negation, as a moment of connection, as a development moment, preserving what is positive' (after Goldzamt and Szwidkowski 1987). According to Lenin's words, historical landscape was divided between the part to be negated and the part to be preserved. Bourgeois housing, religious temples or shrines of the 'opium of the masses', as well as 'useless and reactionist' symbols of old, capitalistic societies had to be eliminated from the cultural landscape, either physically or by shifting function and then significance.

History and policy actively interacted on the cultural landscape scene. Many symbols of 'wrong history' were destroyed. Old monuments representing pre-war governments were among first to be eliminated. In Germany, the 'culture of destruction' was aimed at eliminating most symbolic features of the Third Reich, like Hitler's Chancellery, some ministry buildings, and some other constructions, considered 'not useful', like the few synagogues, department stores and churches in the eastern zone of the city. Since much had been eliminated in Poland and to some extent Czechoslovakia by the Nazis in 1939, the most typical cases, and best analysis, can be found in Hungary (see Fowkes 2002). Decisions reached about the destruction, construction and conservation of political shrines were given careful thought and attention. In the immediate post-World War II period, the new

Communist government faced choices about which monuments to restore, which to remove, and whether to build new shrines. Whether damaged or undamaged, politically unacceptable monuments, such as those to monarchs, were removed (Foote, Tóth and Arvay 2000). A similar process happened in every new socialist country. Not only monuments were to be eliminated; some buildings were seen as very strong symbols of 'negative development'. The list begins with Berlin City Castle, together with Leipzig University Paulinenenkiche (see Chapter 5, Civic landscape discourse) and Budapest Magna Domina Hungarorum Church. There are also numerous houses, palaces and churches that shared the fate of being 'wrong' symbols and were in consequence destroyed. The symbolic demolition was either quiet or done at night, as with most of the monuments, or explained by demands of economic development and infrastructure improvements.

In terms of heritage there was also a positive side to the balance sheet in the communist stewardship of the past. It resulted in much attention being paid to post-war reconstruction of damaged and even completely razed historic cities, most notably in Poland, where the 'Polish Conservation School' became renowned for its state sponsored, thorough, historical reconstructions (Ashworth and Tunbridge 1999). Manipulation of history and memory was popular practice in 'historical rebuilding' of old cities. 'Socialist in substance and national in form' projects were selectively chosen to represent the parts of national history and heritages to meet the current political demand (Nawratek 2005). Socialist regimes, controlling every media and information, often manipulated social memories and imagination. Manipulation and censorship were implemented in many fields, including cultural and heritage landscapes (Mezga 1998). The reconstruction of Warsaw Old Town, destroyed by Nazis in 1944, was undertaken in 1953,[22] used national sentiments, and was done to show the good will of the Party. The Old Town was rebuilt according to late 18th century paintings by Canaletto, while all 19th century transformations were ignored as 'ideologically malignant'.[23] Similar rules were implemented in rebuilding other Polish cities, like Gdańsk or Wrocław, where any traces of German tradition were ignored. The early 1970s decision to reconstruct Warsaw's Royal Castle (Figure 3.8) was to make Polish society to believe that the communist regime was 'our, Polish, national'[24] (see Kołodziejczyk 2007). The Castle, built from scratch, now houses a museum and representation halls, and together with the new Old Town, is one of the main landscape symbols of the Polish capital.

---

22 The year of Stalin's death.

23 In 1980 Warsaw Old Town, constructed in the late 1950s and early 1960s, and was put on a list of the UNESCO World Heritage Sites.

24 During the discussion before the final decision, some of the apparatchiks opted for reconstructing the castle bigger than it actually was by adding an additional floor, to prove that the socialist country can do everything bigger and better, and at the same time hiding view of old town churches (Mezga 1998). Another idea was to rebuild the Old Town and the Royal Castle in the suburban town of Otwock, as an open-air museum (Kołodziejczyk 2007).

Figure 3.8 Reconstruction for the people: Royal Castle, Warsaw, 2007

Historical reinterpretation was also focused on functional transformation. Many older buildings were reinterpreted by implementing new functions. Palaces and aristocratic residences were transformed into museums or recreation houses; some churches were turned into concert halls, storage or museums. Similar fate of functional transition, reinterpretation and in result elimination faced many features typical of capitalist economies and societies. Since services, including any kinds of consumption, alongside financial and market institutions, banks, stock exchanges, department stores, hotels and leisure centres were less necessary in a planned economy, the former establishments were transformed into offices, storage or even flats. Many of Wrocław's old department stores served as offices or storage, reflecting the anti-consumptionist attitude of the communist rulers. The old, large Hermann Tietz's department store in central Chemnitz was transformed into public services, including a library, museum and other cultural establishments.

An important part of historical landscape policy was the negation of any anti-Soviet, anti-Russian or any historical events that could put Soviets in a bad light. The Polish-Soviet war of 1920 was never mentioned or commemorated. Since 1956 in Hungary and Poland and since the late 1950s in the other countries of the Bloc, the Stalinist era was euphemistically called 'a period mistakes and distortions', Joseph Stalin was totally eliminated from the history and landscape, while the school books had to be re-written.[25] The official oblivion aspect of politically facilitated landscape was much broader and incorporated any reminders of Polish, Czechoslovakian and Romanian territories lost to Soviet Union and German history in Western and Northern Poland and Bohemia or Hungarian lost territories. Commemorating any of the forbidden features, events or landscapes was generally interpreted as 'revisionism' and prosecuted.

Communist historical policy was most often opportunistic and was guilty of much neglect, especially in the period of economic stringency, and particularly when the heritage was a reminder of previous royal, capitalist or religious allegiances. Historical policy was often deliberately implemented by not doing anything. Elimination of unwanted landscapes was then limited to leaving old, historical buildings to slow decay. Hundreds of palaces and castles, monasteries, churches, and above all city houses were practically left without owners, or owners were deprived of financial recourses to renovate and modernize the stock. The historical centres had been derelict, and resettled by lower social classes. This gradual but constant destruction can be called *Havanization*, since the Cuban capital is probably the best-known example of the process. It usually took a few dozen years until the buildings were impossible to restore and renovate. Since there was no land rent, there was no market pressure, and unwanted landscape features were left to decay and eventually be forgotten.

Non-socialist features of socialist cities

Regardless of totalitarian state apparatus, not every aspect of social life was fully controlled and facilitated by the omnipotent Party. There were some enclaves of not fully socialist cultural landscape in each of the countries of Central Europe. There were two types of non-socialist landscapes within socialist states: some of them were related to the anti-communist resistance and then for communists considered treasonous and criminal, while other features and activities were merely accepted and regarded as a 'necessary evil'.

Resistance against Soviet and communist domination had accompanied most of the post-war period in many countries of the region and was definitely not accepted by the communist forces. The protests were organized by armed guerrilla groups and farmers threatened by collectivization, but most importantly, both economically and ideologically, were workers' protests, in whose name the communists exercised power. In 1953 strikes paralyzed the tobacco industries of Plovdiv in Bulgaria, and later the Škoda factory in the Czech town of Pilzeň. The same year saw similar outbreaks in the Hungarian steelworks in Cspel outside Budapest, as well as in Ozd and Diósgyör. In June 1953 came the great outbreak in Berlin. In the end of June 1956 strikes in Cegielski factory in Poznań, Poland, developed into street fights and major riots. Autumn 1956 faced probably the largest anti-communist revolution in Hungary, similar to the 1968 Czechoslovakian spring, both of which were throttled

---

25 During my school years in communist Poland I never heard about Stalin, only recalling the streets of the Heroes of Stalingrad. Even Goldzamt and Szwidkowski in their 1987 book on socialist architecture never mention Stalin's name, as if he never existed.

by 'allied' Soviet and Warsaw Pact armies. In December 1970 shipyard workers of Gdańsk, Gdynia, Elbląg and Szczecin struck and marched through the cities. Other protests include 1976 strikes in Radom, Poland; 1977 in Jiu, Romania; and the Solidarity movement started in 1980 in Gdańsk, Poland (Cramtpon and Crampton 2002). Ironically, the protests were not raised against political repression or lack of democracy or freedom, but against the implementation of the economic programme, followed by rising labour norms and food prices. Places commemorating victims of the uprisings were kept in social and individual memories for decades, often just by flowers laid anonymously[26] (see Judt 2005, Hitchcock 2004).

Another 'forbidden landscape' was related to illegal cultural events, usually not controlled by the state. Theatres, libraries, indexed film projection, art exhibitions and concerts had been organized in many private houses and church facilities in major cultural centres of Czechoslovakia, East Germany and Poland during the 1970s and 1980s. Dozens of artists were forbidden from performing publicly after supporting Prague's Spring in 1968, including later president Vaclav Havel. After the military governor introduced martial law in Poland in December 1981 hundreds of artists refused to support the regime and withdrew from the state supported institutions. Information about the events were often distributed only between friends and friends of friends, but some performances were visited by hundreds of anti-communist intelligentsia and students; many others were infiltrated by secret police, while participants were later arrested.

Authorized landscape was often contested and the official places and spaces were reinterpreted. Irony and humour helped to habitualize dogmatic and ideological features. It was a collective reaction against communist propaganda, while jokes and comical stories were aimed to humanize brutal landscapes, as well as ridicule the party and its apparatchiks. Often witty names were given to the prominent buildings and places to reduce the officially given significance. Many of the ideologically significant icons were de-contextualized by contesting societies and de-scarified, at least by some members of the society. There are hundreds of examples of how irony and humour diminished and naturalized alien and often hated symbols. East Berlin's most representative 'Palace of the Republic' was decorated with hundreds of round lamps lined in the foyer and soon earned the nickname *Erichlampenlade* or 'Eric's lamp store', as a humorous tribute to East German leader Erich Honecker. The eagles on the monument to the Polish Army in the Berlin suburb of Hohen-Neuendorf were popularly called the 'Polish broilers', while the huge sculpture of Soviet soldier holding a mall boy in Koszalin, Northern Poland, was known as 'the kidnapper'. 'Osiedle Zasłużonych' ('Quarter on the Meritorious') was built in the 1980s in the up-market district of Wilanów, Warsaw, and was predominantly inhabited by communist party apparatchiks, including the last communist dictator, General Wojciech Jaruzelski. Since the very beginning, the quarter was popularly and ironically called 'the Bay of Red Pigs' in affronting reference to its inhabitants.

---

26 One of those places was Wenceslas Square in Prague, the location where in 1969 Czech student Jan Palach committed suicide by self-immolation as a political protest against the Soviet-led invasion of Czechoslovakia in August 1968.

Private entrepreneurship created some accepted but barely appreciated economic, social and landscape features of socialist countries. In some countries, mainly in Hungary and Poland, private businesses played a very important role in centrally planned economies prone to shortages, offering often better quality goods and services. In some sectors, like greenhouse flowers and vegetable production, bakery, pensions and bars, private entrepreneurs, often called 'craftsmen', substantially supplemented inefficient state industries. On the other hand, private business, all of them only small or medium in size, generated significant personal income,[27] and the new rich could be seen in trendy restaurants and spa centres.

Special roles in the cultural landscape of late socialist cities were played by 'internal export' shops, like *Pewex* in Poland, *Intershop* in East Germany or Czechoslovakian *Tuzex*, where much desired western goods could be purchased paying in US dollars, West German marks or special coupons. The shops were extremely expensive for ordinary citizens,[28] but were also a makeshift of western shops. Those 'oases' of market economy were strictly controlled by the state and only available in the biggest towns and cities. Oases of almost free market and a semi-Western, better world were often connected with the grey sector of economy, illegal currency exchange, smuggling and speculation. The free market sometimes could be seen literally, as in the case of the famous, and largest in the Bloc, the Różycki Bazaar in Warsaw.

Meeting points with Westerners were another important and much appreciated aspect of the cultural landscapes of socialist cities. Since it was generally difficult and very expensive to travel abroad, and especially behind the Iron Curtain, the next best thing was to meet foreigners in socialist country. They usually brought some 'Western glamour', colour, fashion and much desired exchangeable currency. Exclusive hotels, spa towns, airports and the best restaurants visited by foreigners became posh places to see and be seen, 'windows to the world'. A small number of tourists and businessman were hardly visible except in 1970s and 1980s Hungary. In other countries foreigners and all those who wanted to 'touch the West' were attracted by major cultural or economic events, like the international *Intervision*[29] song contests in Sopot, Poland; the film festival in Karlovy Vary in Czechoslovakia; the Chopin festival in Warsaw; or at international fairs in Leipzig, Poznań or Brno.

Religious institution, often deeply infiltrated by securities and secret agents, remained important oases of limited freedom. Churches and religious services not only brought some hope for believers, but were also 'open windows in the closed house' (Hertle and Wolle 2004, 264). Churches were often publishers of the books that could not be published in state-own company, they created an opportunity to work in Christian organizations and to study in the only non-state managed university

---

27 Sometimes a hefty part of the income was illegal, as it was the only possibility to manage an economic activity.

28 In the late 1970s the unofficial exchange rate of 20 USD was the equivalent of average teacher's monthly salary. At the same time it was the price of a pair of Levi's blue jeans in the official exchangeable currency shop in Poland.

29 *Intervision* was the union of Central and Eastern European state televisions, and an answer to the Western *Eurovision*.

east of the Elbe: the Catholic University of Lublin, Poland (see Figure 3.9). The significance of spiritual landscape was most clearly visible in overwhelmingly Catholic Poland, where even many members of the Party regularly and openly participated in religious celebration. The Polish Catholic Church, with support from the West, was additionally strengthened by the Polish Pope John Paul II, elected in 1978. The Pope and his pilgrimages played a very important role in anti-communist resistance. The church, never dependent on the communist state, became a shelter and gathering place for all, not only for Catholics, during the marital law period in the early 1980s. Protestant churches also played a vital role in East Germany, where mass protests and manifestations of autumn 1989 were facilitated by earlier celebrations in churches of Halle, Leipzig and Berlin.

Figure 3.9 Not really a communist feature: Catholic University of Lublin, 2007

It is important to remember that all of the above-mentioned non-socialist features were not principally non-socialist, but were rather abbreviations of market, religious and cultural landscapes, which made a difficult life a little bit easier and answered the de-humanization of 'humanistic' projects of communism. These features were not ideologically socialist, but were usually controlled and to some extend facilitated by secret police and other governmental bodies.

Process of socialist landscape development

The process of landscaping socialist cities can be seen in three main phases. The early Stalinist period was characterized by rapid industrialization, concentrated on heavy industries and post-war rebuilding. The triumphalist landscape of that period articulated the victory of the newly established system, as well as deep belief in the better world it was supposed to represent (Simons 1993, Tarachanow and Kawtaradse 1992, Wegenaar 2004). Post-Stalinist liberalization introduced a second phase of socializing the cities: the International Style and constructivist blocks. The new fascination quickly became the new imperative. The landscape became somehow less ideological and more functional in the 1970s. The invincible and more abstract decision-makers were hidden behind planning desks, central offices and regulations. The last and eventual stage came as a consequence of the failure of modernist project and inefficient economies. The 1980s found the communist system quite corrupt and hardly functional in most of the socialist counties, especially in Poland, where more individual landscape concepts had been implemented (see Leach 1999, Czepczyński 2005a, Judt 2005, Czepczyński 2006c).

Meaningful landscapes in socialist times were always split between the official landscape of the new cult, representative and affirmative buildings and settings, and the scenery of everyday misery, as well as oppression and terror. The system was facilitated by hundreds of labour or slave camps, like Auid in Romanian Transylvania, mines in Dorog in Hungary, Jáchymov in Czechoslovakia, or former Nazi camps Buchenwald, Sachsenhausen, Torgau in East Germany and Łambinowice in Poland. This landscape of terror, assisted by thousands of prisons, helped to usher the new, but not voluntarily chosen system (Crampton and Crampton 2002). Practical elimination of ownership rights ensured the state/Party had the only real power over the landscape, while full control over state owned media monopolized official and well propagated interpretations. Ad hoc and off-the-cuff steering was the most common practice, while the permanent fear and admiration of Soviet models made socialist landscape either pompous and pathetic or repetitive and insincere.

Stalinist-era urban landscaping

From the end of Second World War until the very late 1940s communist regimes in Central Europe were not fully strengthened and consolidated. There were many fields not fully dominated or controlled by the communist party, including shops, small and medium sized enterprises, farms, the press, schools and churches. The first years of landscape and architectural development continued the 1930s modernist tendencies in urban development, especially in Poland and Czechoslovakia. After 1949, when the communists had seized all the political and economic power in Central Europe, a new era of landscape development began, characterized by totalitarianism, repression, and a strong belief in centralized and omnipotent planning as the main tool of modernization in the ruled societies (Simons 1993). Cultural landscapes, as the major component of social and economic life, were strictly restrained and administrated by the communist party. Landscape management was a practical visualization of the victory of socialism and Stalinist ideas and practices. The repetitive model of heavy industrialization, nationalization and overall control by the Party was practically implemented in all Central European countries before 1950. The triumph of communism over capitalism could be best visualized in 'grand designs', but also in repression over the defeated system and its cultural landscape. The conversion also affected cultural landscapes, both in form, by architectural shift from modernism to soc-realism, and in meaning, by signifying any major new construction as 'socialist', 'revolutionary' or 'triumphal'. New urban populations, moved from small rural societies, had to be comforted by 'new, but understandable and close ideas, behaviour and cultural patters; the socio-cultural vacuum caused by migrations had to be filled by positive, educational essences' (Goldzamt and Szwindkowski 1987, 147). Clear, communicative landscapes, were to express progress in national development as well as bright perspectives.

The 1950s triumphalist designs reminded to very much extend the 19th century Haussmannian reconstruction of Paris. Three main goals were achieved by the grand boulevards; technical, economic, as well as anti-revolutionary: wide streets enabled construction of barricades, and made it possible to use regular army troops and artilleries. Correspondingly, Berlin's grand and wide Stalin Avenue, as well as Lenin Avenues in Kraków and Ostrava, were designed not only to frame grand marches, parades and official manifestations, but were also broad enough to introduce tanks and troops.[30] Second Empire Paris was key to urban planning and social control, and the lesson was well learnt by communist decision makers and architects and implemented in many of the early 1950s new towns (Nawratek 2005). The new grand boulevards were supposed to be constructed for 'the mighty human rivers, never seen before in history', as declared by leading Soviet architect Edmund Goldzamt (see Kołodziejczyk 2007). As officially interpreted, the popular demand for understandable, national architecture, easily incorporated by masses of new rural migrants influenced the designers and architects to design neo-pseudo-historical, soc-realistic developments. The transformation was then 'a complex ideological and cultural process, which came from sociological and psychological circumstances' (Goldzamt and Szwindkowski 1987, 146).

---

30 'just in case ...'.

Box 3.5 Largo, Sofia

The Largo is an architectural ensemble in central Sofia, designed and built in the 1950s with the intention of becoming the city's new representative centre. The ensemble consists of the former Party House, now used by the National Assembly of Bulgaria, in the centre, and two side edifices: one today accommodating the TZUM department store and the Council of Ministers of Bulgaria and another that is today occupied by the President's Office, the Sheraton Sofia Hotel Balkan and the Ministry of Education. A Council of Ministers of Bulgaria decree was published in 1951 regarding the construction of the Largo. The lot in the centre of the city, damaged by the bombing of Sofia in World War II, was cleared in the autumn of 1952. The Party House building, once crowned by a red star on a pole, was designed by a team under architect Petso Zlatev and completed in 1955. The Ministry of Electrification office, later occupied by the State Council and today by the President's Office, the work of Petso Zlatev, Petar Zagorski and other architects, was finished the following year, while the TZUM part of the edifice, designed by a team under Kosta Nikolov, followed in 1957. The Largo also once featured a statue of Vladimir Lenin, which was later demolished and replaced by the one of St Sophia in 2000. The cobblestone square around which the ensemble is centred is called Nezavisimost (Independence) Square. It consists of two lanes with a lawn in the middle, where today the flags of all NATO member states stand. Following the democratic changes after 1989, the symbols of communism in the decoration of the Largo were removed, with the most symbolic act being the removing of the red star on a pole atop the former Party House using a helicopter and its substitution by the flag of Bulgaria. In the 1990s there were suggestions to reshape the former Party House, sometimes regarded as an imposing remnant of a past ideology, by introducing more modern architectural elements. According to the new architectural plan of Sofia, Nezavisimost Square is being reorganized and glass lid on the floor was added, so that the ruins of the ancient Thracian and Roman city of Serdica can be exposed, thus becoming a tourist attraction (Largo Sofia 2007).

Stalinist architecture began in 1930s Soviet Russia, and was not based on synthesis or a new interpretation but was highly historical, and focused on detail and décor. Everything had to be 'more' and grander than before (Nawratek 2005). Moscow became a perfect model for a socialist city, so many projects and contractions followed the model of the grandest socialist city. Stalin's Empire, Socialist Classicism or soc-realism came into being in 1933, with Boris Iofan's draft for Palace of Soviets. This architecture resulted from the way the state communicated with the masses through its constructions, using them as an expression of state power. The combination of striking parade monumentalism, patriotic art decoration and traditional motifs has become one of the most vivid examples of the Soviet contribution to architecture. Every point in a design had to gain the approval of the state. Criticism of a project could vary from minor recommendations to total disapproval. As a result many had to be modelled and remodelled many times. This also had a direct effect on the architects themselves, many of whom would later describe this period not by what was actually built, but on what they were not allowed to include.[31] Socialist realism held that successful art depicted and glorified the proletariat's struggle toward socialist progress. It demands of the artist the truthful, historically concrete representation of reality in its revolutionary development. Moreover, the truthfulness and historical concreteness of the artistic representation of reality must be linked with the task of ideological transformation and education of workers in the spirit of socialism. Its purpose was to elevate the common worker, whether factory or agricultural, by presenting his life, work, and recreation as admirable. In other words, its goal was to educate the people in the goals and meaning of communism. The ultimate aim was to create what Lenin called 'an entirely new type of human being': New Soviet Man. Stalin described the practitioners of socialist realism as 'engineers of souls'. The 'realism' part is important. Soviet art at this time aimed to depict the worker as he truly was, carrying his tools. The proletariat was at the centre of communist ideals; hence, his life was a worthy subject for study. The new avenues and wide streets were designed not to be overcrowded by cars, as in the 'rotten West', but for massive manifestations of the working class (see Cohen 1995, Wegenaar 2004).

Joseph Stalin's personal taste, prejudiced by his education and milieu, completely overwhelmed East and Central European architecture and urban landscape in the late 1940s and early 1950s. His admiration of pompous building, kitsch and neo-classical decoration, marbles and crystals, columns and sculptures were copied by numerous followers in almost every country in the Eastern Bloc. Vague Stalinesque copies of Moscow Lomonosov University building appeared in Riga, Prague, Warsaw, Bucharest, Sofia and many other cities (Tarachanow and Kawtaradse 1992). It was officially declared that the Soviet urban tradition of the 1930s had been a very useful source of inspiration and stimulation. It might have been just inspiration, but most likely the new aesthetics had been enforced by regional vassals, or the 'little Stalins', like Bierut in Poland, Gottwald in Czechoslovakia or Ulbrich in East Germany, following the master's thoughts in every little aspects, including urban designs and taste.[32]

---

31 For example, the floral motifs of Art Nouveau were generally not allowed.

32 Today, arguably the only country still focused on these aesthetic principles is North Korea.

Box 3.6 Poruba, Ostrava

The western district of Ostrava, Czech Republic is among the most representative examples of a Stalinist new town. The construction of the town began in 1951, and in 1957 the settlement was incorporated to the town of Ostrava. It consists of a sizable complex of buildings form the 1950s and 1960s inhabited by about 75,000 people. Socialistic history was supposed to symbolize a new workers' society. The architects tried to mirror the history, especially renaissance and baroque styles by sgraffito, sculptures and other ornaments. Many houses were decorated with new types of coats of arms, symbolizing craftsmanship and professions. The district was never fully completed; the plan included, for example, the Oder–Danube canal, modelled on Moskva River. The core, Hlavní Třída (Main Avenue, formerly Lenin Street) is a wide, impressive, classic *Magistrale*, lined with tall houses, Ionian columns, arcades and huge gates. Some houses, like the ones on Porubská Street, not only look like Renaissance castles, but are decorated with soc-realistic cupids, equipped with industrial accessories, like ladles and hammers. The complex is supplemented by the grand crescent on Námesti 9 května (9 May Sq[33]). Interestingly, a 1983 tourist map and guide of Ostrava do not mention Proruba at all, while the 2006 publication lists Poruba as one of the main tourist attractions of the city, since the district was declared an urban historical protected zone on 1 September 2003 (Kremzerowá 2006).

The incorporation of Soviet style was imperative on the agenda of every national architects' association, but inspired and sometimes steered by politburos. For example in 1949 the meeting of the Polish architects formulated its main design recommendations, which soon became paradigms: 'elimination of design styles ideologically reverse and alien to nation, like de-urbanism, formalism, constructivism and functionalism, contrasting of "tight economism", and reaching to Polish traditions and heritages'[34] (Nawratek 2005, 98). Two main values of Stalinist cities, beauty and functionalism, had been realized in traditional-looking streets and squares, designed to 'focus common access. The spaces had been decorated by surrounding efficient buildings, ornamented by façade adornments' (Kotarbiński 1967, 48). This pseudo-conservative architecture was only a clever trick: traditional-looking landscapes were to create an urban illusion of 'living like a bourgeois', but it was only an illusion: the flats were not owned by their inhabitants, and quality and size were far from its models. At the same time many of the artistic details (sculptures, mosaics and paintings) showed an adoration of the new ideology. The general outlook of a building needed to express victory over 'rotten capitalism' and glory of social/Soviet ideas, to arouse a feeling of persistence and power. Extravagant triumphalist exteriors corresponded to the lavish interiors, which can still be seen in

---

33 9 May was the official Victory Day in every Eastern Bloc country, to commemorate the victory over Nazis in 1945. During the communist era all of the Eastern Bloc countries celebrated the capitulation of German to Soviets and the end of World War II on 9 May. Now, like the rest of Europe, we celebrate V-Day on 8 May.

34 New times and new systems required new language. The Orwellian technocratic and pseudo-scientific 'new speak' was very popular in any official speeches and printings, while Orwell's works were forbidden until 1989.

some of the buildings of the period, including the Sheraton Hotel in Sofia, Crown Plaza in Prague (see Figure 3.10) and the first major Stalinist building in Germany – the Soviet embassy in Unter den Linden.[35]

Figure 3.10 Iconic form: Hotel International, now Crowne Plaza, Prague, 2007

The stylistic 'innovations' were accompanied by number of monuments, aimed at becoming focal points of the new urban establishments, by adding comprehensible texts to the settings. The grandest monuments had been raised to Stalin while he was still alive. Hundreds of other important figures were raised into 'altars' and shrines, often released by pre-communist 'dimensioned' heroes. At the same time a personality cult was spread around the region, even grander then Hitler's. Local leaders competed to please the Soviet masters and renamed streets, squares, cities and mountains with the name of Stalin or his allies. There were five Stalin cities in Central Europe, including Stalinogród (Katowice) in Poland, Stalinstadt (Eisenhüttenstadt) in Germany, Sztálinváros (Dunaújváros) in Hungary, Stalin (Varna) in Bulgaria and

---

35 In 1952 the Soviet Union constructed a new building for the embassy in most representational street of Unter den Linden in East Berlin as the representation of its country, but also as a symbol of power and dominance and as an architectural monument to itself. The imposing building with two side wings is three times the size of the pre-war embassy. Long hallways and function rooms are decorated with glass mosaics, exquisite fabrics, marbles, woodwork, mirrors and lush carpets (Kopleck 2006).

Oraşul Stalin (Braşov) in Romania. The highest mountain of Carpathia was renamed Stalinov štít (Stalin Peak, before 1949 and after 1961 – Gerlachov Peak), while several important plans carried J.V. Stalin's name (see Crampton and Crampton 2002).

Box 3.7 Stalin monument, Prague

In 1949 the Czechoslovakian government decided to commemorate the 70th birthday of Joseph V. Stalin by erecting the world's largest Stalin monument. The competition was won by sculptor and architect Jiří and Vlasta Štursa. The largest group sculpture in Europe during its existence, the monument had a reinforced-concrete structure faced with 235 granite blocks, weighing 17,000 tonnes and was 15.5 metres in height and 22 metres in length. A 15-metre-tall Stalin was accompanied by two workers with banners, an agro-biologist, partisan, peasant women, and a scientist, followed by Czechoslovakian and Soviet soldiers. The grand granite ensemble was eventually erected on Prague's Letná Hill in 1955 as an opposition to the old castle. This imposing structure and especially its size aspired to 'squash the past' (Szczygieł 2006, 73). In connection with Soviet criticism of Stalin's 'cult of personality' the monument was 'destroyed with dignity' with 800 kilograms of explosives six years later. In 1990, pirate radio station Radio Stalin operated from a bomb shelter beneath the statue's plinth. The same shelter was also the home of Prague's first rock club in the early 1990s. Since 1991 the marble pedestal has been used as the base of a giant kinetic sculpture of a metronome, the work of sculptor Vratislav Novák, erected in 1991. In 1996 the pedestal was briefly used as a base for a 10.5-metre tall statue of Michael Jackson as a promotional stunt for the start of his HIStory European tour (Asiedu 2005). The City of Prague is considering several options for redevelopment of the site, including a plan to build an aquarium. The empty square, occasionally used by a group of young skateboarders, dominates over the Vltava River, while the Czechs would prefer to not remember the shameful landmark (Szczygieł 2006, 82).

There were three main and par excellence capital implementations of the 'educational' and 'traditional' tendencies in urban design in Central Europe of the early 1950s: Marszałkowska Dzielnica Mieszkaniowa with Plac Konstytucji in Warsaw designed by Józef Sigalin and team; the Largo Sofia with Lenin Square, designed by Piotr Tashev and others; and Stalinalle complex in Berlin. All three establishments have many similarities; all of them have symmetric squares, creating isolated central spaces, free of traffic, but also hardly accessible for pedestrians. Symmetric and monumental buildings accent the main axes of the squares. All of them are based on historical stylization, from classic–modernist synthesis in Berlin, modernist neo-classicism on Warsaw's Plac Konstytucji to monumental heavy rustic with arcades on Lenin Square in Sofia (Goldzamt and Szwindkowski 1987, 156). Similar tendencies can be seen in many other new towns, created in the late 1940s and early 1950s together with construction of big steel works, like Nowa Huta in Poland, Stalinstadt/Eisenhüttenstadt in Germany, Sztálinváros/Dunaújváros in Hungary, artificial fertilizer factories, for example in Dimitrovgrad in Bulgaria or deglomeration of old industrial complexes, like Poruba and Havířov in Northern

Czechoslovakia or Tychy in southern Poland. Stalinist style in landscaping continued for a few years after the death of Stalin, and was first abandoned in Poland and Czechoslovakia, while in Bulgaria and Romania it still continued until the early 1960s.

Supremacy of the International Style

In 1954, a year after Stalin's death, the Soviet All-Union of Workers of Construction and Architecture decided to choose new solutions in urban design, following constructivism and functionalism in architecture. New towns should be based on grand blocks of flats estates, while the whole design and construction process should be maximally standardized. The Central Committee of the Communist Party of Soviet Union published the resolution '*On removal of surfeit decoration in design and construction*' (Szyszkina 1981). A switch from Stalinist architecture towards standard prefabricated concrete is associated with Khruschev's Secret Speech *On the Personality Cult and its Consequences* in February 1956. More applicative and surprisingly earlier than the *Secret Speech*, was the November 1955 decree *On liquidation of excesses in design and construction*. In result of this decree, many 'excesses' had been eliminated or at least minimized, including Stalin's personal cult, some economic and social experiments, as well as liquidation of the landscape and architectural excesses.[36] The new 'quasi-pragmatic period of socialist architecture had begun' and was quickly followed by every Central European country (Nawratek 2005, 78).

The de-Stalinization of the cultural landscape was an important, but rather discreet process. Since the Party was declared to be omnipotent, omniscient and infallible, it was hardly possible to admit mistakes and misjudgements. The inter-socialist iconoclasm had to be inconspicuous but total. History books were re-written, and Stalin's name together with the 'period of excesses' were successfully erased from memories, texts and cultural landscape. Cities, streets, squares, factories, schools were renamed, Stalin pictures went to warehouses, while monuments were melted or furiously destroyed, like the one in Budapest. The Hungarian Revolt shocked Eastern European communist leaders, forcing most to enact economic reforms. The reforms placed more emphasis on producing consumer goods, eased up on farm collectivization, and even allowed some private free enterprise. Rather primitive regimes of early socialism had been replaced by what Foucault called technocratic 'capillary totalitarism' (1986). Certain liberalization of socialist planned economies, as well as cultural and aesthetic alleviation, allowed architects and designers to return to 1930s and late 1940s trends, which were by that time well developed in the West. The process of return to the old-new aesthetics was quite immediate in

---

36 Industrialization of the construction processes was, as Goldzamt and Szwindkowski (1987) implied, the main reason for an imperative shift in urban development in socialist countries after 1956. Officially only the new technologies, not the political shift and de-Stalinization of economies, politics and artistic expressions, were most important reasons to abandon neo-classical triumphal landscapes for modernist free arrangement of blocks.

Czechoslovakia, Poland and Hungary, while in Romania and Bulgaria took few years, and was accompanied by a search for a new, modern but national architecture.[37]

New technologies helped to develop new tendencies in construction and architecture. The industrial architecture developed methods of standardization, using steel, glass, concrete and reinforced concrete, which led to the new architectural style – constuctivism, with the unification of form and construction (see Figure 3.11). The new technique was characterized by simplification of form and elimination of ornamentation. Directly from the attainments of constuctivism came functionalism, where utility is the only aspect that defines the form of the building; with stress on mass prefabrication, optimal use of the construction materials and strict functional segregation. The main ideas of the new style were generally accepted by many, if not most, architects of the 20th century. Two of them – Charles-Éduard Jeanneret-Gris, called Le Corbusier and Walter Gropius, the founder of Bauhaus school of construction, had probably the greatest influence on the style of the time. Corbusier and Gropius were reputed as fathers of the International Style, the radical, functional and constructive school of architecture, based on the rectangular block and extensive access to space, sun and green. Le Corbusier tried to create a 'machine for living', a repeatable and perfect house, which could be built anywhere and everywhere. Extreme and revolutionary landscape concepts went further. 'We must eliminate the suburbs', recommended Le Corbusier. In the new kind of cities, the pleasures of town would be available to all. 'Our streets no longer work. Streets are obsolete notion. Everything here [in the capitalist city] is paradox and disorder: individual liberty destroying collective liberty. Lack of discipline'. The new city would be an arena of green space, clean air, ample accommodation and flowers – and not just for few, but, as the radiant City promised, for all of us (de Botton 2007, 242–245). The revolutionary landscape ideas of Le Corbusier were realized all over the world, although in communist countries, unlimited by market or private investors, those concepts were materialized in their most extreme forms.

The cultural heritage of modernism is memorialized by the most popular form of the time – the tri-dimensional block. The architectural form of block originates from industrial architecture, and follows cost minimization and maximization of production. Block is a massive, homogenic, autonomous structure, dominated by parallel and perpendicular lines of walls and windows. Flat planes of block and static configuration are supposed to generate the unconcern, trustworthiness and steadiness feelings of its inhabitants. The predominant straight line symbolizes order, materialism and nature's subordination to the human being. The homogenious and autonomous form of block impels the 'free spatial structure' of the estates, where the block layout is unconnected to the street pattern. Often none of the walls of the building were parallel to the road. That open plan led to dissipation of the population density and decreased the intensity of usage of the city space. The new block

---

37 As Goldzamt and Szwindkowski (1987, 168) admitted, Yugoslavia 'coming from different historical conditions, did not face such a dramatic transformations'. This could mean that Yugoslavia under Marshal Tito had never been 'Stalinized', and the soc-realistic model had not been forced in Belgrade, so, in consequence, after 1956, the landscape and planning did not go through such a major transformation.

estates required vast areas, usually available outside the city core. The modern city transportation system was designed to minimize the distance from the city centre. These revolutionary changes in the urban planning and architecture find its written form in The Charter of Athens, prepared by CIAM (International Congers of Modern Architecture), under the influence of Le Corbusier. The popularity of the block form of housing was especially high in centrally planned economies, with its mighty, public investors; often municipalities and, particularly in communist countries, housing co-operatives or enterprises. Additionally the metaphoric connotation of straight line, as declared by Le Corbusier, together with egalitarian character of a block appealed to many socialists, not only in the East. The block became a symbol of its times, quite a permanent imprint of *Zietgeist* of the second half of the 20th century and sole and dominant architectural style for housing and institutional building in every socialist country (Basista 2001).

Figure 3.11 Constructivist similarities: blocks in Miskolc, 2006

Box 3.8 Panelák

Very popular housing blocks were given common names, so as to accommodate the inhuman feature into language, and later in to everyday culture. Germans called the blocks *Plattenbau* (*Platte* is slab and *Bau* means building). Probably the most popular name is *panelák,* the colloquial name for blocks of high-rise panel buildings in the Czech Republic and Slovakia constructed of pre-fabricated, pre-stressed concrete. The full name – *panelový dom* (Slovak)/*panelový dům* (Czech), literally means 'panel house'. Buildings remain a towering, highly visible reminder of the modernist and socialist era. In every Central European country *panelák* is a popular form of housing, due to the rapid development of housing constructions during the 1960s and 1970s. It is estimated that about 40 per cent of the total urban population of the region still live in *paneláks*. Similar buildings were built in all communist countries, from East Germany to North Korea. *Paneláks* resulted from two main factors: the housing shortage and the ideology of communist leaders. Planners from the Communist era wanted to provide large quantities of affordable housing and to slash costs by employing uniform designs over the whole country. They also sought to foster a 'collectivistic nature' in the people. Additionally, in case of war (which was always considered), these buildings would not be as susceptible to firebombing as traditional, densely packed buildings. Some of them are more than 100 metres long, and some are more than 20 stories high. Some even have openings for cars and pedestrians to pass through. Many people criticize them for low design quality, mind-numbing appearance, second-rate construction materials and shoddy construction practices. In 1990, Václav Havel, then president of Czechoslovakia, called *paneláks* 'undignified rabbit pens, slated for liquidation'. *Panelák* housing estates as a whole are said to be mere bedroom communities with few conveniences and even less character. In large cities, most *paneláks* were built within *panelák* housing estates. Such developments now dominate the peripheries and sometimes even city centres of almost every city, town and many villages of the region. The town of Most, Czech Republic, is known for having a dominant share of people living in *paneláks*, with about 80 per cent of total urban population living in blocks (Stankova 1992).

Modernity in mid 20th century socialist architecture became official urban paradigm, as were earlier 'classicist' projects. Without much discussion, new models of urban landscaping replaced the Stalinist version of cities. Flat roofs, functional blocks, and the free arrangements of buildings dominated both urban and to some extent also rural landscapes in most Central European countries. Communist urban landscape had been dominated by modernism, Corbousierian constructivism, functionalism and plainness. Blocks of flats were seen as a democratic form of urbanization and city lifestyle, while simplified access for technical and social infrastructure cut the construction costs. Flats in blocks had also higher standard then most late 19th century constructions. Repeatable and simplistic buildings dominated Central European urban landscape, covering all districts with dozen of regular and square blocks. From the late 1950s until the end of the 1970s hardly any iconic or symbolic buildings had been produced, and very few memorable or significant landscape anchors or geosymbols. Anti-symbolic modernism did not produce many visual, mental, social or cultural identification symbols. Grand scale blocks-of-flats estates and some office towers did not create sufficient base for visual place identity building. The concept of block supplied egalitarian and equal housing for everybody and facilitated classless neighbourhoods, but also enhanced anonymity, lack of personal identification with home, lack of neighbourhood relation and watch, and haphazardness of neighbours. Monofunctional residential estates became merely a 'city's bedroom', and were often designed in inhuman and exaggerated scale, with very basic, if any, social infrastructure, while the lack of semi-public spaces impeded integration. Blocks were habitually located in a certain clear geometric order but this was visible often only from a bird's eye view.[38] The marcrospatial scale of the blocks of flats estates strengthened the feeling of totalization of all aspects of existence, including the most private, housing. Flats were characterized by their often small size and limited functionality, collective ownership, and grey, monotone colours. Large districts were usually distant from the city centre, and lack of effective transportation solutions isolated them from many aspects of urban life (see Wagenaar and Dings 2004).

The cities had been totally transformed, since the state/Party held all the assets. The architectural competition 'Socialist Transformation of the Capital of the German Democratic Republic' was a good example of that process. The competition resulted in many projects and designs, planned to re-model the city, change its former capitalist character and prepare for new challenges and new society. Many building arose around *Alexanderplatz*, like the 1969 Television Tower, 1964 House of the Teachers, and many others. Similar transformation plans were applied in central Warsaw, Gdańsk, Lepzig, Dresden, Halle, Varna, Miscolc, Ploieşti and many other cities. Many high streets had been re-modelled in the same style: two-storey shops and service blocks along the street were supplemented by tall housing or office blocks, placed upright to the street. This pattern can be now seen on main shopping streets, like Św. Marcin in Poznań in Poland, or Breiter Weg in East German Magdeburg.

Design offices in most of the socialist countries had been incorporated into large 'construction unions' to rationalize the planning process. The direct effect of the unification was standardization of every aspect of the planning and construction processes, and the very same design of block can be found in many towns and cities around one country.[39] There were several common block designs. For example, the most common series in the GDR was the *P2*, followed later by the *WBS 70*, repeated in hundreds all over the country. The designs were flexible and could be built as towers or rows of apartments of various heights. Most often the blocks were composed in large, more or less elaborated and completed housing estates. The most famous and awarded realizations include Ružinov in Bratislava (designed by Kedro, Pinkalský and others, 1959), Wzgórza Krzesławickie in Kraków (Leonowicz and the team, 1960–1964), Bielany II in Warsaw (Piechotkowie, 1957–1960), Dablice in Prague (Tyček, 1968–1974), Rataje in Poznań (Wellenger and the team, 1950–

---

38 Since this mode of transportation is somehow untypical for most of the users, 'free plan' was often unclear and caused severe disorientation and mislaying. Hexagonal, oval or 'spilled' structures were seen by many inhabitants and visitors only as illogical, just-scattered combinations of same-looking blocks.

39 Which can be seen as realization of Le Corbusier's dream of most effective and everywhere repeatable '*maison citrohan*'.

1952), Nowe Tychy (Wejhert, Adamczewska and the team), Lütten-Klein in Rostock (Urbanski, Lasch and others, 1960–1965), Hipoodram in Sofia (Tashev, 1958–1960) and Tolbukin in Burgas (Siromachov, Kassarov, 1964–1970) (see Goldzamt and Szwidkowski 1987). Since the late 1960s there was a rapid growth of huge blocks of flats housing estates. Those giant settlements were build on the edges of cities, like Marzan and Hochschönhauser for 200,000 people in north-eastern Berlin, Petržalka in Bratislava populated by approximately 130,000 and Ursynów in southern Warsaw for more then 100,000 people.

Box 3.9 Przymorze, Gdańsk

One of the largest in Poland, this model block of flats neighbourhood was located in the Gdańsk district of Przymorze (see Figure 3.12). The estate was designed by Tadeusz Różański, Janusz Marek and Danuta Olędzka and prefabricated by the local 'house factory'. The project, awarded first prize by the Polish Architects Association in 1959, comprised housing districts for 50,000 people on 200 hectares neighbouring the Baltic Sea coast. Public utilities were supposed to cover 70 per cent of the land. The plan included one primary school for every 960 pupils, cinema-theatre complex with 700 seats, vocational schools, shopping centres, sport fields and many others. The average sized flat was only 40 square metres, with 12 square metres per person, according to obligatory housing standards of the times. The concept was based on a series of blocks perpendicular to the coastline, 11 stories long blocks, with characteristic broken lines of the walls, popularly known as *falowiec*, or the waver. The largest of them all is the characteristic block on Obrońców Wybrzeża Street, about 800 metres long, 11 stories tall and designed to house approximately 5,000 people. The main blocks were supplemented by a set of five-storey buildings, vertical to the *falowiec*. The district was divided into four smaller units, called A, B, C and D. The project was developed by a housing cooperative, 'Przymorze', while many of the flats were guaranteed for the shipyard workers. Most of the initial project was realized, except for non-basic services. Due to its convenient costal location, trees, recently developed services and shopping facilities, Przymorze is still an attractive housing district,[40] both for older residents, as well as for newcomers.

Implementation of constructivist landscape policy met many obstacles and barriers. Systems of quality control and supply of components were often insufficient and caused many delays, alterations of the initial projects and very low quality of construction and finishing. Furthermore, there was no clear decision-making and management scheme. Many rulers implemented many various regulations, while the landscape management power was often divided between territorial and sectoral planning and decisions. The over-regulated system created the popular practice of circumvention, obeying the set of laws, making many exceptions, forced or persuaded by grand powers of socialist industry and the Party. Often single high

---

40 In 2007 a small 30 square-metre studio could be bought for about 60,000 euro.

rank landscape lord completely changed the primary deigns[41] (Basista 2001, Leach 1999, Czepczyński 2005, Domański 1997).

Figure 3.12 Megalomania on the coast: Przymorze, Gdańsk, 2006

Late socialist landscape ideas and endeavours

Crises of the concept of international proletariat and global communism appeared from the late 1970s, especially in the Soviet Union and Poland. Even many high ranking party officials and apparatchiks did not believe in official propaganda and the egalitarian 'happiness project'. Forty years of attempts and experiments in Stalinist triumphalism and constructivism had not been able to create a 'new society' or 'new man'. The crises of the ideology became quite obvious in most of the Central European countries. The communist system, especially in its last years, was totally permeated with cynicism, falsehood and vanity. Almost each aspect of its existence was false (Esterázy 2007). The 1980s brought many symptoms of the coming transformation.

---

41 As, apparently, it happened in the new town of Tychy, Poland in the 1970s. The new and powerful secretary of the local communist party had high aspiration for his town, and wanted Tychy to be seen as metropolitan and important. It is said that at one of the meetings he said 'do you want Manhattan? So I will build you a Manhattan'. In consequence the initial, nationally awarded low-rise design was changed to include a number of tall housing towers, which expressed ambitions of the old apparatchik.

The signs of change could be seen on many levels, including political and economic, especially extreme in Poland, but also on the architecture and landscape. The liberalization of economic and political life was mirrored in anti-totalitarian and divertive landscapes, in every socialist state in Central Europe.[42] During the late 1980s, liberalism and self-governance were considered the main methods of limiting the omnipotent state. While liberalism was connected with primacy of individuals and private property over so-called 'social benefit', self-governance was accelerated by primacy of local society, governing via chosen representatives (Nawratek 2005). Many Central European economies, together with the Soviet Union, showed clear symptoms of falling down. The misery of late socialism could be seen on many aspects, also in landscape. Many towns, dominated by modernist architecture, looked similarly miserable, as described by Wendell (2003, 88–89):

> [T]he blocks, eight of ten stories high, as regular as bricks, march one after another in an equidistant parade along the road. They are dead grey – the colour of a leached cloudy winter sky – twenty or thirty years old, pock-marked with crumbling concrete, scabbed plaster, rough and peeling paint. Some grey god put them there, squashed foundations against the dry land and lifted concrete boxes one on the top of another. The same god who brought people down from the mountains and prised them from their villages to work in the factories towards the great Industrial Future. (...) The windows in the blocks are square, the balconies strips of narrow reinforced cement stacked like columns of vertical dominoes. Over the years, people have walled in the balconies with a rough collection of breeze-blocks, red-brick, tiles, corrugated iron, plastic sheeting and widow frames, to make an extra room. The over effect is patched and patchwork; small efforts of DIY individualism.

Bankruptcy of modernist projects was accompanied by a number of other landscape solutions. New concepts had to be developed to replace corrupt, disappointing and failed ideas of international workers' egalitarianism. Usually, new concepts were inspired or installed by the Soviet Union, but from the late 1970s Soviet leaders were not able to provide any acceptable solution, and the overwhelming process of dissolution was also visible in Moscow – the communist focal and reference point. 'Interim themes' of nationalism, historicism and limited individualism were among the last communist envisages, somehow preparing societies for an inevitable and unthinkable end. So-called 'socialism with human face' turned out to be the last, evidently not human enough, type of socialism in Central Europe. Societies and Party leaders were looking for new-old identities, and the most popular were historicisms and nationalisms. The turn was visualized in the preservation of historical cities, like Kraków and Zamość in Poland, Sibiu and Braşov in Romania and Quedlinburg in Germany.

The interpretation of history had been changed, and past prides and achievements had been incorporated into national and social heritages. Assimilation of historical landscape was focused on rather distant and usually medieval history. Renovation of historical buildings was accompanied by the rise of monuments commemorating important national events. Probably the largest shift towards historicism and tradition

---

42 But also in Soviet Union.

was seen in Bulgaria, together with the celebration of the 1,300[th] anniversary of Bulgarian statehood in 1981. Many monuments were raised to commemorate this event, with the largest one in the country dedicated to the Creators of the Bulgarian State dominating the town of Shumen. Reminiscence of historical landscape was also visible in new urban designs. Free planning was replaced by traditional streets, while flat roofs and square block were substituted by more historical-looking houses. The process of 'historization' of cultural landscape began in the late 1970s, and was especially popular in Poland and East Germany.

Box 3.10 Edelplatte

East German historicism was connected with the acceptance of its own history, and resulted not only in the renovation of some historical and almost abandoned towns and castles, but also in brand new, but historically inspired projects. The 1980s witnessed more decorative and historically inspired prefabricated houses, decorated with mosaics, stucco, loggias and corner balconies to better integrate with historical surroundings. The construction of Nikolaiviertel in central Berlin was aimed at re-creating the 'old town' as new a tourist attraction. Prefabricated segments were used to shape 'medieval' looking houses, which eventually gave a rather peculiar landscape. A similar effect was achieved on another Berlin historical square, Gendarmenmarkt, where again, modern materials were used to construct 'stylized' buildings, reminiscent of the original 18th century architecture. Every time, the reference period was carefully chosen and accepted by the high political officials, since the copied forms carried more then just outlook, but the historical reference too. Rebuilt 'medivalish' or 'classicistic' houses were related to the promotion of the times of Brandenburg electors and Frederick the Great. In 1985 the former government district of Wilhelmstrasse, almost completely destroyed during the war or a few years after, yielded to the last GDR mega-apartment complex in the inner city. Apartments in the *Edelplatte* (noble prefabs) were available only to most privileged and selected people due to their relatively high build standard and vicinity to the Wall (Kopleck 2006). The very last landscape projects[43] brought rather a surprising historical reference. The houses on Otto-Grotewohl Street, former Wilhelmstrasse, although prefabricated and grey, were generally based on the German late 19th century, or *Wilhelminian* architecture (see Figure 3.13). The reference to a recently very much hated capitalistic period might indicate deeper political and social transformation during the last years of German Democratic Republic.

Another path of landscape development is connected with folklore, widely present in the urban cultural landscape of socialist cities. Rural tradition was always an important part of communist workers' and peasants' propaganda, especially in countries where actual workers were in sharp minority. Rustic and country heritages had been promoted in number of open-air museums, state sponsored and often inspired folk dance groups, revitalization of traditional country clothes, habits and cuisine, which became important tourist attractions. 'Folklorite' restaurants were a must for a foreign tourist, while regional and rustic inspired souvenir products

---

43 Most of which was actually completed after the fall of the Wall in 1990 and 1991.

were the most popular, and often the only advisable, souvenirs from Eastern Bloc countries. Folklore heritage was also employed in political indoctrination, as a sign of achievements of rural proletariat. Folklorism was particularly strong in Bulgaria and Romania, but also in Poland and Hungary, which can be seen in state published travel guides of the time (see Paszkowiak and Pelzer 1976, Dreichfuß 1987).

Figure 3.13 German prefabricated historicism: neo-Wilhelminian blocks, Berlin, 2003

Developments in landscape and leisure were also new signs of change in late communist Central Europe. In the 1970s and 1980s many new resorts had been constructed, especially on the Black and Baltic sea coasts, but also on Lake Balaton and the Czechoslovakian mountains. The most popular sea resort was probably Zlate Piaski (Goden Sands) in Bulgaria. The spa town, sometimes called the 'socialist St. Tropez', but more like Costa del Sol, was the socialist tourists' dream, and a subsidized alternative for the unreachable Mediterranean. Mass tourism also resulted in new hotel developments in major urban centres. The 1970s faced a limited, but larger then ever, flow of foreign investment, accompanied by high standard hotels. The most expensive and exclusive were visited mainly by foreign businessmen and the most privileged locals, while others could only see them in films and TV programmes. Socialist luxury hotels had to be exclusively Western, and actually mainly belonged to Intercontinental Hotels, as in Warsaw, Sofia, Prague, Budapest and Bucharest.

Development of consumption facilities, including large department stores, addressed growing standards of living. Sizeable facilities appeared in every bigger city, including the Zentrum network in GDR, Domy Towarowe Centrum in Poland, Kotva in Prague or Unirea in Bucharest. Interestingly, one of the visible signs of 'new' landscape was exclusive, and 'almost Western looking' orange mirrored glass. Since the 1970s orange glass decorated many representative buildings, like Belrin's Palace of the Republic, apartment towers and offices on Váci út in Budapest, House of Culture in Prague or Polish Airline office in Gdańsk.

The Romanian leader Nicolae Ceaușescu chose a more original way of landscaping cities and re-urbanization. The 'grand constructor', also known as 'the sun of Carpathia', initiated the project, based on radical 'systematization' – many new buildings were built in previously historical areas, which were razed and then built upon from scratch. A major part of Bucharest's architecture was made during the late communist era with 'more efficient' high-density buildings. From the late 19th Bucharest was sometimes compared with Paris. The 'Parisian complex' and ambition to be even bigger and better, can be seen in the neo-Haussmannian and neo-Stalinist design of the Civic Centre (see Figure 3.14).

Figure 3.14 Stalinist revival: Civic Centre, Bucharest, 2005

Box 3.11 Civic Centre, Bucharest

One of the best examples of late socialist architecture is *Centrul Civic* (Civic Centre), a development that replaced a major part of Bucharest's historic city centre with giant, often ten-storey tall utilitarian buildings, mainly with marble or travertine façades, inspired by Stalinist architecture. Eight square kilometres in the historic centre of Bucharest was levelled and 40,000 people were resettled in order to make way for the grandiose *Centrul Civic* and the immense Palace of the People (Figure 4.7). This complex of modern concrete buildings is centred on a boulevard originally known as the Boulevard of the Victory of Socialism, renamed after the Romanian Revolution of 1989 as Unification (Unirii) Boulevard. The Boulevard, modelled after Paris's Champs-Élysées and a few metres longer than the original, runs roughly east–west, constituting a grand approach to the Palace of the People at its western terminus. *Centrul Civic* includes numerous government offices and apartments, the latter being roughly equal in number to the housing units destroyed for its construction. The apartments were originally intended to house Romania's communist elite, but the completed complex is certainly not a preferred residence for the city's new capitalist elite, with the possible exception of buildings that look out on the now-bustling Unirea Square, where *Centrul Civic* bisects the Dâmbovița River, which is channelled underground past the Square. *Centrul Civic* stands out through its high degree of architectural uniformity, but also through its lack of commercial space. The vast empty fields which emerged in the historic town during the demolitions of the 1980s were sarcastically called *Ceaușima* (a portmanteau of Ceaușescu and Hiroshima). Many buildings of the *Centrul Civic* have never been completed, like classicist National Library and Academia Romana (Centrul Civic 2007, Ioan 2007).

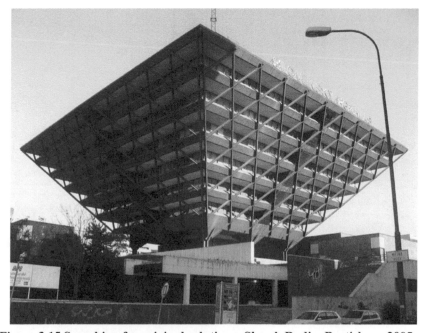

Figure 3.15 Searching for original solutions: Slovak Radio, Bratislava, 2005

The last years of socialist rule over Central Europe brought some release and certain liberation in economic, social and architectural endeavours. Some of the counter modernist landscape projects, especially those designed by Czechoslovakian architects, as can be seen as peculiar architectural prophecies. Structures like the nine-storey upside-down pyramid of Slovak Radio (Figure 3.15) or the concrete brow of the National Gallery in Bratislava, together with the glass brick tank of the new National Theatre in Prague give rise to feelings of uncertainty, ambiguity and confusion in most passers-by. Some of those 'prophetic structures' seem to say 'the end in near', and in fact, it was.

Socialist landscape was socialist not only because it was constructed during the communist era and under the auspice of the communist rulers. Communist era economic and social management practices and implementations implied total control over practically every means of production, media and education. The absolute hegemony triggered totalization of landscape to an extent very seldom seen in any other developed countries. With all the command in the hand of the omnipotent Party, anything and everything could have been built, and any dream could have came true, no matter costs and sacrifices. But this landscape was socialist mainly because of the ideological texts attached to almost every project. Virtually everything was or was supposed to be socialist in meaning. When landscape features were considered 'non-socialist', then in a similar manner to people, institutions and ideas, they became anti-socialist and passively or actively eliminated from public view and memory. There was very little room, if any, for neutrality in social life, language, culture and cultural landscape under socialism. During each of the analyzed periods, forms, functions and meanings reflected national hopes and aspirations, powers and fears, together with ambitions and limitations.

Chapter 4

Post-communist Landscape Cleansing

The transformation of urban landscape can be evolutional, as in most cases, slowly and steadily changing urban features. Landscape evolution follows new philosophical and aesthetic ideas, where new forms progressively replace the old ones. The other, revolutionary form of landscape development reflects a rapid and usually complete conversion of political and iconographical systems. Landscape revolution goes behind the political transformation, and can be as fatal and turbulent as any revolution can be. Landscape then becomes a battlefield, where buildings and arrangements representing opposing ideas become enemies and rivals, as well as victims and winners (see Czepczyński 2005a, Sármány-Parsons 1998). The problem of how to manage inherited, received or leftover landscape is common in every transitional society. The transformation of meaning, use and attitude to the obtained landscape was very often similar for post-imperial, post-capitalist and now post-socialist cities. The codes and meanings written into the landscape still reflect the past values and ideas that represent the inglorious precedents. The need for reinterpretation and transformation of repressive principles of former rulers follow political, social and economic transformations (Leach 1999). More generally, the question is raised as to whether a past should and could be publicly ignored. There are many arguments in favour of an official policy of collective amnesia. It may aid recovery from past trauma and also permit the healing of social divisions, especially when those who benefited from, and those who suffered under the old regime must coexist in the new system. Against this is the argument that it has never proved possible in practice to eradicate a past through coercion in the long term: it tends to return at some future date, as has been the experience of a number of Western European countries with their Second World War heritage (Tunbridge and Ashworth 1996).

After socialism, cities faced vast legal, economic and social conversions. Changes have been accelerated by the explosion of free market and flow of capital, reintroduction of land rent, privatization, as well as the appearance of new actors on the landscape, including local governments, free media, private owners and investors, as well as inhabitants and non-government organizations (NGOs) (Czepczyński 2005a). The post-socialist 'landscape swap' is most clearly visible in large cities and metropolises. The accumulation of needs, capital and powers made this swap most dramatic in sizeable urban settings. Many smaller towns and villages are characterized by generally more stable scenery, slowly interacting with the grand but distant events and powers. Post-socialist resolutions of the early 1990s were characterized by a fairly spontaneous understanding of freedom on both personal and institutional levels. After more than 40 years of oppression and restraint the control mechanisms almost disappeared. Control over the landscape has been mostly seceded to local self-governing bodies, frequently somewhat unprepared for the new challenges and responsibilities.

For most participants in the post-socialist discourse, many landscape meanings and signs are closely related and deeply anchored in the pre-1989 realities and relations. This re-production of meaning in the urban landscape has been happening for almost 20 years, and is definitely not completed: a considerable part of the contemporary Central European system of representation still represents and signifies concepts, signs and objects of the communist past. Every revolution or deep transformation movement pays lots of attention to the symbols, destroys some of the monuments and on their locations builds its own, because the revolution wants to fix and petrify itself, both materially and metaphorically – to survive (Kapuściński 1982). Obsessive and complex symbols and icons seem to be essential for successful revolutions, since those visible signs speak to everybody on its victory and envisage position, power and hope. Revolution destroys and changes, but never fully, utterly, peremptorily. Usually the former system created and left immense material and cultural resources, so the new, often less strong system is not able to obliterate or eliminate them completely. Sometimes the new system is interested in leaving the marks of the old one (Kapuściński 2007). It is usually easier to change political system than the mentality and beliefs of the local society. Cultural landscape, as every cultural product, stays longer in minds and hearts of the practitioners. Ordering and rearranging cultural landscape always becomes one of the most important tasks of the new landscape lords. Cleansing or preserving a landscape feature is an act of historical policy and represents local, regional and national needs and expectations.

Liminal times – liminal landscapes

Communist rule in Central Europe was directly dependant on Soviet political, economic and military pressure and support. The Soviet Union intervened militarily in Hungary in 1956 and again in Czechoslovakia in 1968. Even under this threat, the societies of Central Europe increasingly complained about their lack of political freedom and the failure of socialism to increase their standard of living. Starting in the early 1970s, Polish workers joined food riots and called strikes that led to the formation of Solidarity, a nation-wide pro-democracy movement. During the mid 1980s, a younger generation of Soviet apparatchiks, led by Mikhail Gorbachev, began advocating fundamental reform in order to reverse years of Brezhnev stagnation and period of severe economic decline. The first signs of major reform came in 1986 when Gorbachev launched a policy of *glasnost* (openness) in the Soviet Union, and emphasized the need for economic reform – *perestroika* (restructuring). Gorbachev urged his Central European counterparts to imitate *perestroika* and *glasnost* in their own countries. However, while reformists in Hungary and Poland were emboldened by the force of liberalization spreading from East to West, other Eastern Bloc countries remained openly sceptical and demonstrated aversion to the reform. These regimes owed their creation and continued survival to Soviet-style authoritarianism, backed by Soviet military power and subsidies. Believing Gorbachev's reform initiatives would be short-lived, orthodox Communist rulers like East Germany's

Erich Honecker, Bulgaria's Todor Zhivkov, and Czechoslovakia's Gustáv Husák obstinately ignored the calls for change.[1] By the late 1980s, it became clear that the Soviet Union would no longer use its military to keep the Central European communist parties in power. People had lost all faith that the communist system could deliver a better way of life. In 1989, people everywhere in the region took to the streets and overturned the communist regimes one after another. In a matter of months, the system imposed on the countries of Central Europe by Stalin for 40 years disappeared (see Judt 2005, Simmons 1993, Sorin and Tismaneanu 2000).

The political upheaval began in Poland in 1980, when the free trade union Solidarity (*Solidarność*) was legalized. December 1981 martial law stopped the transformation for a while. The collapse of the communist system was accelerated in the late 1980s, when the nationwide strikes in 1988 forced the government to open a dialogue with Solidarity and form a 'round table' negotiation scheme, completed by the first, partly free elections of 4 June 1989. The total victory of Solidarity resulted in a new non-communist government. The elected parliament designed a new, non-communist government, the first of its kind in the region, which was sworn into office in September 1989. Following Poland's lead, Hungary was next to revert to a non-communist government. In 1989 Parliament adopted a 'democracy package', which included trade union pluralism; freedom of association, assembly, and the press; a new electoral law; and a radical revision of the constitution, among others. In October 1989, Parliament adopted legislation providing for multi-party parliamentary elections and a direct presidential election (Revolutions of 1989 2007).

After Hungary's reformist government opened its borders, a growing number of East Germans began immigrating to West Germany via Hungary's border with Austria. By the end of September 1989, more than 30,000 East Germans had escaped to the West. The mass exodus generated demands within East Germany for political change, and mass demonstrations with eventually hundreds of thousands of people in several cities – particularly in Leipzig – continued to grow. Unable to stem the flow of refugees to the West, the East German authorities eventually caved into public pressure by allowing East German citizens to enter West Berlin and West Germany, via all border points, on 9 November. Hundreds of thousands of people took advantage of the opportunity; new crossing points were opened in the Berlin Wall and along the border with West Germany. The opening of the Berlin Wall proved to be fatal for the GDR. By December the Socialist Unity Party of Germany's monopoly on power had ended, which led to the acceleration of the process of reforms in East Germany that ended with the reunification of East and West Germany that came into force on 3 October 1990. Emboldened by events in neighbouring East Germany, and the absence of any Soviet reaction, Czechs and Slovaks rallied in the streets to demand free elections. On 17 November 1989, a peaceful student demonstration in Prague was severely beaten back by the riot police. That event sparked a set of popular demonstrations. By 20 November the number of peaceful protesters assembled in Prague had swelled from 200,000 the day before to an estimated half-million. With other Communist regimes falling all around it, and with growing street protests, the

---

1 'When your neighbour puts up new wallpaper, it doesn't mean you have to too', declared one East German *Politburo* member (Revolutions of 1989 2007).

communist Party of Czechoslovakia announced on November 28 they would give up their monopoly on political power, as a consequence of this non-violent 'velvet revolution'. On 10 December, the Communist leader Gustáv Husák appointed the first largely non-communist government in Czechoslovakia since 1948, and resigned.

Unlike other Central European countries, Romania had never undergone even limited de-Stalinization. In November 1989, Ceaușescu was re-elected for another five years as leader of the Romanian Communist Party, signalling that he intended to ride out the anti-communist uprisings sweeping the rest of Central Europe. Ceaușescu ordered the arrest and exile of a local Hungarian-speaking Calvinist minister, László Tőkés, in the Western Romanian town of Timișoara, on 16 December, for sermons offending the regime. Tőkés was seized, but only after serious rioting erupted. After learning about the incident from Western radio stations, the demonstrations spread. On the morning of 22 December, the Romanian military suddenly changed sides. Army tanks began moving towards the Central Committee building in Bucharest with crowds swarming alongside them. Ceaușescu and his wife, Elena, escaped via a helicopter, while dozens of protesters were massacred outside of the building. On Christmas Day, Romanian television showed the Ceaușescus facing a hasty trial, and then suffering summary execution. An interim National Salvation Front Council took over and announced elections for May 1990. On November 10, 1989 Bulgaria's long-serving leader Todor Zhivkov was ousted by his Politburo. In November 1989 demonstrations on ecological issues were staged in Sofia, and these soon broadened into a general campaign for political reform. In February 1990 the Party voluntarily gave up its claim on power and in June 1990 the first free elections since 1931 were held, won by the moderate wing of the Communist Party, renamed the Bulgarian Socialist Party (see Judt 2005, Simmons 1993, Sorin and Tismaneanu 2000, Revolutions of 1989 2007).

The conversion of powers in Central Europe had been very fast[2] and somehow unexpected. Even Solidarity leaders did not expect to gain all the power within few months. The changes had been happening on many different levels. Transformation of post-socialist countries after 1989 can be classifies in three main types:

- Political: A shift from authoritarian dictatorship towards parliamentarian democracy, based on a coherent legal system, where society was an active participant of the governing processes and procedures. Political instability and frequent transformation of political parties left the electorate somehow lost in multiple choices. Regional policies forced de-centralization of power and recreation of local municipalities.
- Economic: A centrally planned state economy was replaced by one that was private and market oriented, based on free competition of entrepreneurships and liberalization of market rules. Privatization, collapse of old socialist industries and foreign investments changed local economic rules, while unemployment rose as one of the main economic and social problems in most countries of the region.

---

2 A sign seen in Prague summed it up this way: 'Poland – 10 Years; Hungary – 10 Months; East Germany – 10 Weeks; Czechoslovakia – 10 Days'. 'Romania – 10 Hours' was added after the revolution in Romania (see Revolutions of 1989 2007).

- Social: Started by contesting the forced interpretation of the communists' social ideas. The egalitarian imperative parity was replaced by differentiation and pluralism. Civic rights and freedom of thought boosted the rising aspirations. Freedom of movement caused vast migrations, especially after joining the EU, reshaping many local and regional labour markets and societies (see Sorin and Tismaneanu 2000).

All these types or aspects of changes have been always visualized in space as function, feature or meaning, contributing to the cultural landscape phenomenon. Landscape at times of profound and structural transformation represents social and cultural trends and tendencies, sometimes hidden under the layer of declarations and practices. The transformation or liminal state is characterized by ambiguity, openness and indeterminacy. One's sense of identity dissolves to some extent, bringing about disorientation. Liminality is a period of transition, during which our normal limits to thought, self-understanding, and behaviour are relaxed, opening the way to something new. People, signs, places or things may not complete a transition, or a transition between two states may not be fully possible. Those who remain in a state between two other states may become permanently or long-term liminal. Victor Turner (1975) gained notoriety by exploring Arnold van Gennep's (1960) threefold structure of rites of passage and expanding theories on the liminal phase. Van Gennep's structure consisted of a pre-liminal phase (separation), a liminal phase (transition), and a post-liminal phase (reincorporation). Turner noted that in liminality, individuals were 'betwixt and between': they did not belong to the society that they previously were a part of and they were not yet reincorporated into that society. Liminality is a limbo, an ambiguous period characterized by humility, seclusion, testing and haziness (Turner 1975). Those liminal times can be branded by liminal landscapes: landscapes no longer typical of the previous regime and planning, but the same time quite different from the ones aspired to.

Three liminal phases, as explained by Turner (1975), can be clearly seen in contemporary post-socialist cultural landscape transformation. Cultural landscape is, in a sense, a living laboratory of transforming meanings and forms. Regional and spatial differentiations around Central Europe help to observe and analyze phases of liminal alternations:

- Separation is the first phase of liminality, which began just after first free elections in 1989. Sorting out the 'good' and the 'bad', and new definitions and codes were common in this epistemological transformation. Landscape cleansing occurred directly after the process of separation.
- Transition is most typical liminal state, characterized by a mélange of meanings and representations. The old landscape is re-interpreted and de-contextualized, while the new landscape is constructed, both physically and mentally.
- Reincorporation is the final rite, when the division between 'old' and new' becomes insignificant and eventually disappears. This phase may have just begun in Central Europe, and most likely will be implemented by new generation.

Liminal transformation of Central European cultural landscape consists of multiple separations, transitions and reincorporation, expressed in political statements, everyday practices and living spaces. Radical changes in urbanized landscape administration resulted in spatial confusion and a certain level of anarchy. Newly elected self-governments, both at regional and local levels, had to cope with repeatedly changing regulations, as well as high expectations of the local communities. The recently freed inhabitants, released from the ruthless chains of socialist regulation, expected to enjoy the rights of private ownership, to an extent seldom met in Western European countries. The personal taste of new decision-makers, as well as national history, heritage and financial recourses, were mirrored in the features of the emancipated urban landscape of the early 1990s (Leach 1999; Sármány-Parsons 1998). The burdensome meaning of communism was usually left deeply coded into both external and internal structure of urban landscapes. The problem of dealing with meanings and forms of post-socialist leftovers was one of the most significant issues of post-socialist landscape management.

The 1989 'autumn of nations' brought not only overturn of the communist dictatorships, but also the opportunity of finding new paths towards the future, freed from traces of fear, cowardice or renunciation. This future is still very much conditioned by the past and its interpretations. While the functional significance of landscape has always changed, the form has seldom been radically transformed. Combining Hall's (2002) approaches to representation with Turner's (1975) phases of liminality can create an interesting framework to classify and synthesize post-socialist landscape conversions. The process of communist icons' transformation has been categorized in three main types: elimination of reflective icons, transition of intentions and reincorporation of new social constructions to the deep-rooted landscape features.

Separation and elimination of mimetic meanings

Eliminating any unwanted attributes of cultural landscape has been the first step in post-socialist landscape cleansing throughout all Central Europe mainly in the early 1990s. The landscape was cleansed of unsolicited elements and qualities, to make cities more habitable and acceptable for the liberalized societies. Sometimes post-traumatic societies block parts of their collective memory as a remedy to deal with a distressing and hurtful past (Ricoeur 2004). This practice employs the omnipotent mercy of oblivion: features forgotten are not important any more. Since most humans are inclined to keep positive memories and forget the traumatic ones, a hefty part of the former communist landscapes and icons are more or less forgotten by now. Many old icons simply disappeared from public view and people's minds. The leftover landscapes of emptiness or silence, such as empty pedestals, can be meaningful only for those who dare or care to remember. Many of the unwanted codes and symbols, names and labels have been physically eliminated and demolished, especially of features hard to reinterpret, followed by their elimination from social practices and memories.

A central part of the transition is based on the rejection of many aspects of the 'recent past'. Almost all revolutions begin with the idea of 'year zero': a new beginning founded upon the eradication of what went before. Equally almost all find this collective voluntary amnesia an ultimately untenable position and return either to conciliated versions of old pasts or feel the need to create a new past in support of new identities and aspirations (Ashworth and Tunbridge 1999). All new governing ideologies recast heritage, and communism had left an enormous legacy of public iconography. Removal, renaming, rededication or just reuse of the symbolic heritage of a discredited regime was, in itself, simple enough, 'a new onomatology of places' (Węcławowicz 1997). New names and new celebrated heroes not only symbolize new political and historical references, but are also in opposition to the old, politically, economically and morally bankrupt system. This revolutional transformation of places and landscapes is each time aimed to redefine and reposition past, encourage and enhance the expected patterns of thoughts and behaviors. Memory becomes then a crucial module of this process, since it is always very closely related to spaces: physical, social, semiotic and mental. Manipulation or, more politically correctly phrased, memory management turns out to be unmistakably vital element post-communist policy, oscillating between reminiscence and oblivion. The concept of 'thick line', introduced by the first non-communist Polish Prime Minister Tadeusz Mazowiecki in 1989, was to put the history aside, look forward and together, despite of political differentiations, build the independent country. The 'thick line' has been implemented not only in politics, but also in the interpretation of socialist icons. It seems like a significant portion of post-socialist societies would rather 'put history aside' and not evoke the most painful memories.

Generally speaking, since the early 1990s the political aspect of cultural landscapes in post-socialist cities has begun to disappear. The opening landscape transformation tactic has been based on the reflective or mimetic approach of representation (see Hall 2002), derived from the belief that meaning remains in the objects, places and buildings in the real world, while language functions like a mirror to reflect or imitate the true sense as it already exists. There are some landscape features, like monuments of names, where the meaning seems to be truly located within an entity. A sculpture or street name more or less directly reflects the icon, and many people believe that landscape directly mimics a system of concepts. The elimination of structures and objects thought to be mimetic was most spectacular, theatrical and often most remarkable. The process of purging can be material or mental, and always follows liminal separation of good/acceptable from the offensive/undesirable/unwanted. Separation, the first phase of liminality, began just after the first free elections in 1989 and 1990. Sorting out the 'good' and the 'bad', redefining and re-coding started this epistemological transformation, while 'landscape cleansing' went directly after the process of separation. Political iconoclasm is typical revolutionary behaviour aimed at reconstructing and reinterpreting the past by eliminating unwanted icons that strongly represent the old system. Since 1989 the process has involved renegotiating the meaning of historical events and persons and has affected the way these events have been represented and commemorated in the landscape. After four decades of iconoclastic strategies implemented by the communist parties, a new post-communist iconoclasm has been activated by local governments, associations, political parties and individuals (see Foote, Tóth and Arvay 2000).

Eliminated symbols

Changing and eliminating unwanted features or residua of the political landscapes were among the most demanded and sometimes risky tasks decision-makers and managers. The drive to de-communize public space was particularly strong in Poland, Romania and Hungary, as well as in the Czech Republic. The key role was played by the new right wing, nationalistic and anti-communist parties and governments, which usually anchored their identities in anti-socialist, anti-Soviet and often anti-Russian narratives (see Leach 1999, Sármány-Parsons 1998). Landscape features reflecting communist ideas had to be eliminated from public spaces. Changes and removals made after 1989 were always selective. The question was not whether to remove all the statues put up during the communist regime, nor whether to return all place-names to their pre-communist forms but to eliminate the worst and physically unacceptable icons and oppressive signs of the fallen regimes. Communists erected many apolitical monuments, and assigned numerous inoffensive toponyms. Many of them were left unchanged, but others returned to their historical forms or were changed to honour new heroes. Some statues were removed, others were modified,[3] or restored and reconstructed (Foote, Tóth and Arvay 2000).

One of the first tasks in the elimination of unwanted meanings was the process of selecting and purging emblems, logos and coats of arms. Since 1989 each of the Central European country has modified its socialist emblems and formal representations. Red stars, together with hammers and sickles disappeared, to be replaced by crowns and historical symbols. Sometimes, like in Hungary, East Germany or Romania, the national flag with a hole in place of the socialist logo symbolized the 1989 revolution. The socialist symbols and slogans vanished from shop windows, streets, train stations, houses, factories and even farms. Armies changed uniforms, while internal police forces and some ministries changed names, so as not to be associated with their communist predecessors. Even post-communist parties seldom carry any visual symbolic relation to the past. Former communist and workers' parties became social democratic, and sometimes even the meaningful red colour was replaced by neutral and modern blue, like in the Polish Social Democracy.

The process of eliminating communist cultural symbols could reflect to some extent the speed and depth of social transformation. The stones on the former Central Committee of the Socialist Unity Party (SED) in Berlin were carefully replaced from an upper part of the western façade, so the holes left after removing the grand SED logo could not be traced. The systematic eradication in Germany left hardly any major communist symbols in place. At the same time, however, the coat of arms of the Peoples' Republic of Bulgaria has been barely censored: the sickle and hammer had been roughly chipped off the stony façade of the today's House of the President in Sofia (see Figure 4.1). An interviewed older Bulgarian said that communism in

---

3 Often done only by removing the most controversial red star or symbolic hammer and sickle.

Bulgaria disappeared as much as the Soviet logo from the old coat of arms: just a superficial re-make on the surface, while the socialist merit is still left inside the structure.

Figure 4.1 Marks of communism: former logo on former centre of power, Sofia, 2005

Changing symbols were accompanied by the replacement of geographical names. The political map of Central Europe was fully converted: no single country kept its socialist era name; each dropped the 'People's' or 'Democratic' adjective, and Poland and Czechoslovakia returned to the pre-Second World War form. The pre-war monarchies of Hungary, Romania and Bulgaria became republics, and the former German Democratic Republic was incorporated into the German Federal Republic. Many streets have been renamed from Marx, Lenin, October Revolution or Red Army to a variety of local heroes or historical names. Similar procedures have been applied to reflectively communist towns' names, like Karl-Marx-Stadt (town of Karl Marx) returned to its historical name of Chemnitz in East Germany, Gottwaldov[4] to Zlin in Czech Republic, and Hungarian Leninváros (Lenin's town) to Tiszaújváros (Crampton and Crampton 1996).

In spring 2007 Polish right wing and populist government worked on new regulations, aimed at eliminating any leftist/communist street names still remaining

---

4 Named after 1950s Czechoslovakian communist leader Klemet Gottwald.

in some municipalities. For the Polish president and his twin bother prime minister, communist-associated local names were a shameful and illegal promotion of communist ideas. Prime Minister Jarosław Kaczyński said that 'there are still some monuments in Poland that honour people who committed severe crimes against humanity, against Polish nation; many streets are named after obvious criminals' (Szcześniak 2007, 2). The purge of Marxist iconic names sometimes went rather far, eliminating any technically left-associated names, together with some 19th century social and workers' activists. Karl Marx himself patronizes only a few streets in Central Europe, most of them in former East Germany, and none in Poland. Any communist-related patrons, like Walery Wróblewski, the Polish general fighting during the Paris Commune in 1871 or Ludwik Waryński, who established the Polish Socialist Party 'Proletariat' in 1882 seemed to be 'too mimetically communist' to remain on the pedestals. The officially authorized and organized process of institutionalized iconic landscape revisioning followed 'landscape cleansing' and reflects the current course of political and social actions. As street-naming has been a municipal responsibility, local names with some socialist connotations remained in many of the left-voting towns. Local societies traditionally opting for right-wing, anti-communist or nationalist parties have quite efficiently cleansed their landscape from any 'left-sounding' names. The process is very clearly visible in more traditional Catholic south-eastern Poland, with hardly any place-names reflecting communism, while in the more liberal and social-democratic north-western parts of the country, street and squares still carry names of minor communist activists.

Abolished forms

The fate of monuments of iconographical socialist heroes can rather accurately illustrate political and social transformations and conditions of liminal societies. One of the most common practices in 1989 and the very early 1990s includes physical destruction and demolishing of objects impossible to reinterpret, like much-hated monuments. The statues symbolized malevolence and misfortune, as well as the supremacy of the communist system. The very same practice was common in early years of communist rule, when many of the pre-communist monuments, as well as churches and palaces, were blown up to vacuum the scenery (Fokes 2002). There are three strategies to recontextualize old monumental icons. The most spectacular one was based on the 'remove and destroy' (or, sometimes, destroy and remove) approach. In some cases, the process of icons' removal became a festival and symbolic gesture of liberation. The Berlin Wall (Figure 4.2) became the most popular icon reflecting the division of Europe, as well as of communist supremacy and isolation.

Figure 4.2 Reconstructed symbol of the Cold War: Berlin Wall, 2004

Box 4.1 The Berlin Wall

The Berlin Wall (*Berliner Mauer*), known in the Socialist Bloc as the 'Anti-Fascist Protective Rampart', was a separation barrier between West and East Germany. An iconic symbol of the Cold War, the wall divided East and West Berlin for 28 years, from the day of construction on August 13, 1961 until it was dismantled in 1989. The Wall was over 155 kilometres long. In June 1962, work started on a second parallel fence up to 91 metres, further into East German territory, with houses in between the fences torn down and their inhabitants relocated. A no-man's land was created between the two barriers, which became widely known as the 'death strip'. During this period 125 people were killed trying to escape to the West, according to official figures. However, a victims' group claims that at least 1,245 people had been killed trying to flee East Germany. Newly discovered documents confirm that the communist regime gave explicit orders to shoot to kill attempted defectors, including children. When the East German government announced on 9 November 1989, after several weeks of civil unrest, that entering West Berlin would be permitted, crowds of East Germans climbed onto and crossed the Wall, joined by West Germans on the other side in a celebratory atmosphere. Over the next few weeks, parts of the wall were chipped away by a euphoric public and by souvenir hunters; industrial equipment was later used to remove the rest of it. The fall of the Berlin Wall paved the way for German reunification, which was formally concluded on 3 October 1990. Little is left of the Wall at its original site, which was destroyed almost everywhere. There are three sections still standing: an 80-metre piece near Potsdamer Platz; a longer section along the Spree River near the Oberbaumbrücke nicknamed East Side Gallery; and a third section in the north at Bernauer Straße, which was turned into a memorial in 1999. None still accurately represent the Wall's original appearance. They are badly damaged by souvenir seekers, and fragments of the Wall both with and without certificates of authenticity are a staple on the online auction service eBay as well as German souvenir shops. Moreover, the eastern side is covered in graffiti that did not exist while the Wall was guarded by the armed soldiers of East Germany. Previously, graffiti appeared only on the western side. In October 2004 an illegal memorial and 200-metre long copy of the Wall was build by a private museum close to famous Checkpoint Charlie on Friedrichstrasse, although not in the location of the original Wall. Under pressure from local developers and plot owners, accompanied by social protests, the copy of the Wall was demolished in July 2005 (Berlin Wall 2007, Klopeck 2005).

A good example of the cleansing of local icons is the removal of the statue of Felix Dzierżyński[5] from Warsaw's Dzierżyński Sq. (now and before the War: Bank Sq.) The relegation of the bronze sculpture of the much hated 'Bloody Felix' was accompanied by enthusiastic crowds, singing, drinking champagne[6] and celebrating symbolic 'breaking the chains' in autumn 1989. Some kept a piece of the smashed hero as a souvenir of gone-for-good history. The remains of the monument are stored by the municipal gardening company in the outskirts of the city (Dudek 2005). Several of the old icons in bronze were melted to make material for new statues, or were sold to private collectors, like Kraków's Lenin to Italy and Berlin's one to Holland. Some other seems to be 'disappeared and forgotten'; including Sofia's Lenin statue removed in the late 1990s due to a road reconstruction and never returned to its former place or the Bucharest Lenin's monument, moved from its high pedestal in front of the 'House of Free Press' and laid down by the kitchen wall of suburban palace of Mogoşai, visited only by foreign tourist equipped with the Rough Guide (see Figure 4.3).

Red Army memorials were usually monumental structures and played a significant political function during the communist era as symbols of dependence on the Soviet Empire. Tanks, obelisks and grand sculptures of victorious soldiers have evoked many bad memories. Those monuments were often located in central parts of the cities, major crossroads or hills, so the local society was reminded every day of to whom they should be thankful. Many of the Soviet war memorials were removed from the most exposed and central locations after 1989. Those in cemeteries were maintained and protected according to the international conventions and treaties governing war graves. Most spectacular Soviet Army monuments and memorials were in the capital cities, like the grand complex in Berlin Treptow or Liberation Statue on Budapest Gellért Hill. There is no question that the Red Army suffered high casualties liberating Central European countries from the Nazis, but the goodwill engendered by liberation was expended many times over during

---

5 Also known as 'Iron Felix', a member of the Polish gentry and later founder of the much hated first Soviet State Security *Cheka* in 1917 (All-Russia Extraordinary Commission to Combat Counter-Revolution and Sabotage).

6 Ironically the very popular (and only available) champagne in the Eastern Bloc was the Russian equivalent of champagne – *Shampanskoie Igriiskoie.*

four decades of Soviet domination that followed. Public war memorials were easy targets for removal from central to peripheral locations as soon as the Soviet Army withdrew, while the remains discovered were moved to the military cemeteries[7] (see Foote, Tóth and Arvay 2000).

Figure 4.3 Junked icons: Lenin monument in Mogoşai, 2005

Relatively few statues and monuments were actually destroyed after 1989. This means that some very powerful emblems of the communist period remain in public view, albeit somewhat off the beaten track. In every country of the region many statues were literally re-positioned and de-pedestaled and removed to peripheral locations, but the Hungarian cases are best documented and analyzed (see Foote, Tóth and Arvay 2000). The most famous and the biggest is Szobor Park in Budapest (see Figure 4.4).

---

7 The Red Army monument, locally known as a 'Bronze Soldier', was erected in the city centre of Tallinn, Estonia, in 1947 to commemorate the capture of the city by the Soviet Army. Until 1991 it was the focal point of Soviet celebration in Estonia. The conflict rose when the Estonian government decided to remove the monument and the remains of nine Soviet soldiers to the military cemetery at the outskirts of the city. The monument became a symbol of war suffering and liberation for the Russian minority and new occupation and Soviet domination of Estonians. The street fights at the end of April 2007 resulted in one Russian dead and several hundred Estonians and Russians injured, as well as many plundered shops and destroyed cars (Harding 2007).

Figure 4.4 Cemetery of old icons: Szobor Statue Park, Budapest, 2005

Box 4.2 Statue Park, Budapest

After the fall of the communist regime in Hungary in 1989, many of the communist statues and monuments were immediately removed. The issue of what to do with all the statues dating from the previous political system was one of many that occupied public debate after the political changes of 1989–1990. On 5 December 1991, the Budapest Assembly came to a decision concerning the future fate of the statues in question – the choice of statues to be removed or kept would be decided by each district individually. The Cultural Committee of the Assembly invited a tender for 'what is to be done with the statues', which in effects was a tender for the design of the future Statue Park. The winner was local architect Ákos Eleőd, who designed the 'Cemetery of Public Monuments of the Recent Past' in Budapest 22nd district of Szobor. In the autumn of 1993, the museum was finally opened, but it was not completely finished. The Statue Park comprises dozens of monuments, statuettes and plates relocated from the streets and squares of the Hungarian capital, including Lenin, Marx and Engels, memorials to the Soviet Soldier, the Communist Martyrs, as well as local Hungarian communist leaders such as Béla Kun and many more. Many, but not all the communist statues of Budapest were gathered together as beacons for the political and ideological culture of the former socialist period (Szobor Park 2007). Visitors to the park mainly comprise of foreign and Hungarian tourists. Now that the great excitement and controversy over the opening of the Statue Park has died down, it looks set to take its place among the many museums and sites of Budapest.

Other 'de-pedestaled' iconic statues are located in Kozłówka, Eastern Poland and is a much modest congregation of Marxist memorabilia, based mostly on 'unwanted icons storage' established during de-Stalinization times just after 1956. Sometimes, the creation of 'monuments' parks' can be accidental and facilitated by local entrepreneurs.[8] A local businessman from Uniejowice in south-western Poland declared: 'I have 10 ha and space for any monument, if somebody has a problem with it'. In the mid 1990s nearby local councils seceded two monuments of Karol Świerczewski, the Polish general who fought against the fascists during Spanish War and was later hero of the Polish People's Army. There are also monuments of Marceli Nowotko, leader of the Polish Communist Party and General Aleksander Zawadzki, together with the Soviet pilots' obelisk from neighbouring Legnica (Przytułek dla pomników PRL 2007). These specific theme parks are mostly visited by tourists, and become just another interesting attraction, but seldom a history lesson. De-sacralization and de-pedestalization of old icons brought them down to the mere position of a tourist attraction, while the iconic and ideological significance has been reduced to their market appeal.

After the initial purges of the early 1990s, there is still certain demand for a 'refuge' or 'asylum' for the old monuments. The process of separating and eliminating statues and has not been completed in Central Europe. Many smaller memorials have been raised from the 1960s until the 1980s, and most of those which remained on their former locations have simply had their inscriptions removed, which can be seen in the town centres of Ostrava and Chemnitz. Monuments without inscriptions became insignificant and practically forgotten relicts of the past. Leftover and text-less monuments become merely curious sculptures, often hidden in unkempt green, like the small statue group on Corvin Street in Miskolc, northern Hungary. Probably sooner or later these derelict monuments will disappear from public view. Another wave of elimination of mimetic icons accompanying local political shifts towards anti-communist right has happened in Poland since 2005. Historical and sometimes 'hysterical' policy creates immense pressure on local governments to purge any post-communist residua still remaining, even sometimes against the will of the local population. In consequence of this historical policy, a stone head of Janek Krasicki, young Polish communist and alleged Soviet spy, was removed from Gdańsk's district of Oliwa to a park in Kozłówka in the summer of 2007. With the implementation of newly proposed Polish acts of law on places of national remembrance, 'statue parks' will become much desired storage places of unwanted, and possibly soon illegal, monuments.

At the same time, very practically, only a few iconic buildings were mimetically communist enough to be destroyed in the course of cultural landscape clearance in Central Europe. Since buildings can be much easier re-defined and reused, only the most important ones had to be destroyed. Only in 1999 was the mausoleum of Georgi Dimitrov, Bulgarian communist leader, torn down in downtown Sofia by the

---

8 In 2001, a Lithuanian anti-communist businessman opened a private theme park known as Stalin World or Grūto Park, 120 kilometres south-west of Vilnius. Two Lenins, Stalin, Brezhnev, Dzierżyński, together with Mother Russia and many more sculptures as well as exhibitions are spread over 20 hectares of forest.

right-wing royalist government. For many, the other victim of revenge and purge is Berlin's Palace of the Republic, called by some vicious commentators the '*Balast der Republik*' (see Figure 4.5). The demolishing began in 2006, officially due to asbestos contamination in the structure of the building, but for many of the East Germans the reason was clearly political: eliminating a prominent symbol of the late German Democratic Republic.

Figure 4.5 Disappearing icons: demolition of the Palace of the Republic, Berlin, February 2007

Box 4.3 The Palace of the Republic, Berlin

The Berlin City Castle was the most significant baroque castle north of Alps and the residence of Prussian kings and emperors. The construction survived the War relatively unscathed. For the communist governors, the edifice, which had evolved since 1443, symbolized the hated Prussian militarism and imperialism. The castle was blown up on 7 September 1950 and vast parade grounds were temporarily formed on the ground on the castle. Designs to build a tall, soc-realistic governmental building[9] were never realized, while a large and centrally located space within the former castle was renamed Marx-Engels Square and used for mass gatherings and manifestations. Between 1964 and 1989 the nearby State of Council Building of the GDR housed the office of the head of the governmental council. The modern building is decorated with a copy of the fourth portal of the City Castle, from which on 9 November 1918 Karl Liebknecht declared

---

9 Similar in form to Warsaw's Palace of Culture and Science.

the German Socialist Republic. In 1972 Erich Honecker procured the construction of the Palace of the Republic, a hall in glass and gold serving many purposes, including the Chamber of People, a powerless parliament of the GDR, 5,000-seats conference hall, as well as a popular restaurant, many thematic clubs, theatre, bowling hall and an art exhibition (Kopleck 2006). Just prior to German reunification in October 1990, this building was found to be contaminated with asbestos and was closed to the public. After reunification, the Berlin city government ordered the removal of the asbestos. In November 2003, the German federal government decided to demolish the building. Notwithstanding many protest of East Berliners, demolition started in February 2006 and was scheduled to be completed in 2007, but the process has been delayed, and the ruins are likely to totally disappear by 2008. The government was publicly accused of modern iconoclasm, eliminating unwanted symbols of the late East German 'miracle'. The eminent orange mirrored-glass building, the pride of modernist design, will be replaced by a copy of the imperial City Castle, which will house the Humboldt Forum, an international forum of art, culture and science. The Forum is designed to act as a site of information, education and cultural encounters for both Berliners and visitors, facilitated by a selection of exhibitions, lectures, theatres, film, music and dance. The Forum 'will unite Wilhelm von Humboldt's humanistic model with Alexander von Humboldt's ideal of all-encompassing knowledge. On the grounds of the former Palace, at the very centre of historic Berlin, a new, universal approach to the world's civilizations will emerge' (Removal of the Palast der Republik 2007).

Empty pedestals and former sites of the monuments, like the one left after the removal of the world's largest Stalin monument in Prague, holes left after memorable plates, vast squares and broad avenues designed for grand marches and meetings, silently speak of 'the recent past'. The message of these landscapes of silence is only understood by those who still remember. Usually, many of the early transformed icons are well forgotten. Fewer and fewer people can remember the old, socialist street names, exact locations of the monuments or the sites of former communist party buildings, not even mentioning meanings and texts officially attached to those icons. The practice of landscape transformation mirrors social expectations and demands. One can notice the somehow limited, incomplete and slow elimination of residual features of socialist cultural landscape in East Germany and Bulgaria, in comparison to Poland or Hungary. That process might indicate different political preferences and attitudes towards the 'recent past'; in eastern Germany the old icons of GDR are ignored, half-forgotten or considered as parts of their national identity, while in history-oriented Poland most of the remembrances of the former People's Republic recall oppression, misery or at least lost opportunities.

Excluded functions

The oblivion procedure incorporates implementation of new vital functions, followed by meanings into the former communist icons. New roles, often contrary to the old ideological ones, cover the former meaning, while frequently leaving the form unchanged. Generally speaking, it is not very difficult to change the intention and function, especially after some years. Since most of the re-codification and reinterpretation happened over 18 years ago, the former, ideological meaning has commonly been forgotten. The process of transforming objectives by fixing new intentions to the old icons can be best exemplified by role of the former communist party buildings in almost every Central European town. Regional or municipal communist party quarters almost everywhere lost their primary intentions and significance. The former centres of power and supremacy were turned into much less dominant, but locally important public buildings like schools, offices, banks or culture centres. One of the classic examples of the alteration of function and meaning is the former headquarter of the Polish United Workers' Party in Warsaw. The 1950s structure, locally known as the 'White House', was reassigned in 1991 as one of the first and the biggest financial centres in the region, the Warsaw Stock Exchange and later as the Warsaw Financial Centre (see Figure 4.6). From an icon of workers' power the building became a new icon of financial and capitalist supremacy. Most of the other communist parties' headquarters faced similar de-classification, from the main source of power to an office building of secondary administration (Berlin, Budapest, Prague, Bratislava). Only in Sofia and Bucharest are the grand classicist buildings of former Communist Party Central Committees still among the most important governmental buildings. Every changing function of former communist iconic building turned out to be quite challenging. Using huge, sometimes uncompleted and mentally connected to the hated system buildings can be difficult task for local and national governments, as can be seen in central Bucharest.

Figure 4.6 Converted icons: former headquarters of the Polish United Workers' Party, now Warsaw Financial Centre, 2004

Box 4.4 The Palace of the Parliament, Bucharest

The world's second largest building was constructed in the centre of the Romanian capital Bucharest (see Figure 4.7). The Palace of the Parliament (*Palatul Parlamentului*) is the central feature of the grand Civic Centre (see Chapter 3, Process of socialist landscape development; Late socialist landscape ideas and endeavours). The classicist, 1920s Bucharest-architecture inspired structure was the expression of the very last Romanian communist leader's megalomania and trend-setting aspirations. Ceauşescu, similarly to Stalin and Hitler, wanted to be remembered as a 'great constructor' and wanted to personally control and approve almost every stage of 'his' project.[10] The structure combines elements and motifs from multiple sources, in an eclectic postmodernist architectural style. The combination of cultural and aesthetic illiteracy, rigid Marxist-Leninist orthodoxy and an innate taste for gigantism somehow represents the uniquely Romanian form of socialism accompanied by similarly exceptional urban landscape. The edifice has 1,100 rooms and is 12 storeys tall, with four additional underground levels currently available and in use. Construction of the Palace began in 1983. The building was originally known as the House of the People (*Casa Poporului*), and sometimes as House of the Republic (*Casa Republicii*), and was intended to serve as headquarters for all the major state institutions. However, the project was just nearing completion at the time of Nicolae Ceauşescu's 1989 overthrow and execution. Since 1997, the building has housed Romania's Chamber of Deputies; the Romanian Senate joined them there in 2005. The Palace also contains a massive array of miscellaneous conference halls and salons. In 2003–2004 a glass annex was built, alongside external elevators to access the National Museum of Contemporary Art opened in 2004 inside the west wing of the Palace of the Parliament, and to the Museum and Park of Totalitarianism and Socialist Realism, also opened in 2004. Parts of the building are yet to be completed (Palace of the Parliament 2007). There are public tours organized in a number of languages, but seldom in Romanian, due to rather limited interest in past. The Palace grew to be one of the main local symbols and tourist attractions of Bucharest (see Ioan 2007, Petrescu 1999).

Grand open public spaces in socialist cities existed as meaningful places only until they were politically used. At all other times, grand squares and avenues were hardly anything more then modes of transportation, just a street or square, or, if peripherally located, did not existed in social memory or consciousness. They were often too big even to walk around. At the beginning of the transformation period, it became clear that large urban areas had to be redeveloped. Many of the places and landscapes became absolutely useless, while new functions and social need put more pressure on redefinition of some of the 'urban fallows'. Landscapes based on meeting the political need of socialist socialites and their powers began to wane. Generally speaking, since the early 1990s the political aspect of cultural landscapes in post-socialist cities began to disappear. Ideological functions of grand open spaces have been replaced by parking or unofficial skate-parks.[11] Vast, empty and

---

10 It is said that he ordered the destruction of the grand marble stairways of the Palace of People twice, because it did not suit his vision.

11 Usually the good-quality granite surfaces of formerly politically important squares are overtaken by skateboarding youth, who typically do not realize and are not interested in the history of 'their' playground.

hardly manageable squares are easily degraded, meaningless, and then often sold, redeveloped and forgotten (Nawratek 2005). In the early 1990s central Warsaw's Square of Defilades (*Plac Defilad*) became the city's largest open-air market, replaced later by a conglomerate of provisory, but still functioning tin market halls. The hall, together with an adjoining car park, will give place in the near future to the new complex of Museum of Contemporary Art. Some of the old places of manifestation have recently regained some of their public function and meaning. Public demonstrations and protests, especially in Budapest and Warsaw, are organized and practised often on the 'historical' squares of communist-era meetings.

Figure 4.7 New interpretations of not too old features: Palace of the Parliament, Bucharest, 2005

Another function seceded to oblivion is connected with the revolutionary cults and shrines of the system. There were hundreds of larger or smaller museums, which were established to educate, indoctrinate and propagate communist ideas all around the communist Central Europe. Most of them had hardly any historical artefacts, but were popular destinations of organized and not fully voluntary tourism in the communist era. Museums of revolutionary or/and workers' movements were probably the most popular type of ideological shrines. Every major city in Bulgaria, Romania and East Germany had to have one. There were also museums personally dedicated to Lenin, like in Warsaw, Kraków or small village of Poronin in Poland, to Marx and Engels in Bucharest, or Rosa Luxemburg and Georgi Dimitrov in Leipzig. None of the old museums exist anymore, and almost nobody remembers the locations. The smaller buildings were by and large returned to the former owners, while some of the larger ones are still public edifice, including museums, but definitely not communism related. Communist propaganda, education and research institutes were among other significant landscape features to disappear from post-socialist, especially capital, cities. Institutions like the Institute of the Basic Marxist–Leninist Problems in Warsaw, the Institute of Czechoslovakian Communist Party History in Prague or the Institute of Marxism–Leninism in Berlin were established to strengthen ideological and intellectual bases of further economic and social developments of 'dictatorships of proletariat'. Even some of the graduates of those institutions do not recall the full name, while the locations are either unused, like in Berlin or transformed into commercial schools and offices.

In 1991, in an artistic act of reinterpretation, Czech artist David Černy painted a Soviet tank-monument in pink, and many monuments were dumped from the pedestals, which were sometimes re-used for other, more politically correct, monuments. Destruction or elimination always had the very same result: empty space is left, and after a while, nobody remembers what was there. Properly descriptive evidence of totalitarianism might be better used than a nice square with pansies. The act of destruction of a monument, a mimetic symbol of the past and reviled power, was a particular act of *catharsis*, a way to start from 'new beginning'.

Transition of iconography: changing intentions

According to Turner (1975), transition is the most typical liminal state, characterized by an unbalanced mélanage of meanings and representations. The old landscape is being re-interpreted and de-contextualized, while the newly constructed scenery answers new intentions and is continuously being constructed, both physically and mentally. The intentional approach of representation is based on belief, that signs and symbols mean what the instigator intends they should mean. The intentions of the investor, designer or decision-maker can sometimes be clearly read in urban text, but sometimes the initial purpose is forgotten or replaced by decisions of the new 'intentioners'. Transition of intention changes the representation and significance of symbols and icons (Hall 2002). New political and cultural rules imply changing codes and purposes of former socialist icons. Re-structured objectives, placed between oblivion and reminiscence, involve re-writing history and re-positioning the past. Each society and individual person should have the time and possibility to choose what they would like to erase from their memory, what to leave and what to remember. Recently, in some countries, including Poland, 'the only right' assessment of the past has been declared by government and political powers. More then 20 years ago a similar appraisal resulted in the withdrawal of titles and names from library shelves and school teaching programs (Hennelowa 2007). Cultural landscape icons of Central Europe, like monuments, places and names are being re-negotiated and re-interpreted, as much as communist past and memories. The changes were sometimes spontaneous, sometimes well planned and designed.

Transformation of intention is the least stable and most turbulent phase of liminal landscape transformation. Changes in interpretation vary from one social group to another, from election to election, from region to region. Negative aspects of the communist iconology are being forgotten or carefully kept and used against political rivals. Landscapes of transition are the result of battles between varieties of versions of the past. Iconic sense has been changing while these signs and objects have been used for far-reaching political activities (Denzer 2005). Immanent aspects of transition are related to multiple *lieux mémoire* or *polyphonic memory* (see Nora 2006, Traba 2006). The same landscape arrangement can be associated with different intentions of various social groups. Simultaneously intentions can be changed, sometimes very fast, as a reaction to the changing stock of information, medial campaigning or shifting priorities and aspirations. This transitional 'landscape in between' remains most difficult to analyze, since multiple and changeable intentions are followed by diverse interpretations in constant revolutionization. The process of transition can be illustrated by numerous landscape icons, like the Palace of Culture and Science in Warsaw.

Box 4.5 The Palace of Culture and Science, Warsaw

The largest and tallest building in Poland was completed in 1955 and presented as a 'gift from the Soviet nation to the Polish nation.'[12] Initially, and only for a year, the building was officially called 'Joseph Stalin Palace of Culture and Science', and still houses the Polish Academy of Sciences, libraries, two museums, three theatres, cinema, congress hall, bars, restaurants, viewing platform and many other cultural and scientific entities. This vast, limestone sky-scraping tower is the fourth largest building in the European Union, and dominates Warsaw's city centre (Figure 2.1). The 231-metre high, 42-floor, over 3,000-room 'wedding-cake-like' structure was designed by Russian architect Lev Rudniev, but apparently Joseph Stalin himself was the author of the concept. The building was constructed by 3,500 Russian workers. Rudniev wanted the skyscraper to incorporate features of the Polish national style. Polish Renaissance architecture was generally regarded as the most revered, so it was nominated as Poland's socialist and national format. In order to come to this decision, Rudniev and his Russian team had toured well-known Polish cities and towns like Kraków and Zamość finding out about Polish architecture and incorporating Renaissance attics and décor into the soc-realistic structure. The building is additionally decorated by numerous sculptures and carves relief symbolizing 'new man' and new times. On New Year's Eve of 2000 the clock was unveiled on top of the Palace. This Millennium Clock is considered the highest tower clock in the world. After long deliberations, since February 2007, the Palace has been listed as historical and national heritage and is legally protected from any reconstruction or transformation of form (based on History of PKiN in a nutshell 2007). Vivid medial discourse and diverse reaction to protecting this 'symbol of Soviet occupation' or 'one of the most interesting buildings in modern Warsaw' still speak of importance of the old intentions and present interpretations (Passent 2004). Newspapers' reactions on the listing of the Palace reflected multiple and diverse attitudes towards the edifice and, more generally, the communist past. The opinions varied from relief and pride: 'Palace eventually listed. Small, but tasteful' (leftist and post-communist *Trybuna*) to outrage: 'Stalin's gift and symbol of Soviet Empire became historical monument' (*Dziennik*). For many the Palace is still a symbol of Soviet domination and communist tyranny. Interpretations are sometimes bold, as that of Jan S. Stanisławski, who declared that 'Palace has to go! Something being so shamelessly erected in the middle of the city must have obscene connotation'. At the same time, especially for younger generation, the Palace, actually older then Warsaw's Old Town, is the coolest and most funky symbol of the Polish capital city.

---

12 Not that we really wanted this Palace, but nobody was asked.

Another characteristic of post-socialist transitional landscape is represented by *Large Abandoned Objects* (Wendell 2003). Leftovers and half-forgotten residua represent misfortunes and mistakes in central planning, economic crises and inconsequence, but also giantomania and the too-high aspirations of former decision-makers. Uncompleted, unused and derelict half-bridges, half-roads, half-built towers or just derelict and empty houses, factories and offices, where everything sellable has been stolen, can be still seen in many cities across Central Europe. The *Large Abandoned Objects* are quite common in countries like Poland and Romania, where deep economic crises suddenly stopped any further development. Buildings like Bucharest House of Radio, National Library or Academia Romania remind the viewer of old aspirations and bad or unrealistic planning. In Kraków, the Main Technical Organization building, the over-scaled structure locally known as the 'skeleton', has stood uncompleted for more then 20 years. Wendell (2003) describes this transitional landscape to the point:

> Rusting tractors, bits of pipelines, lines of coal cars shunted and left along a rail line, half-built bridges and apartments blocks standing concrete and empty; skeletal, burnt out crane hanging over them. Bits of past left, ruined collective farm barracks and factories with all their windows smashed and their plant ripped out of scrap. Remnants of something else – a civilization of sorts? – scattered everywhere, lumps of concrete and bits of twisted metal, lying about to stub on your toe. Everywhere lays the debris of the Soviets, the husk of an empire. Mostly where were hundreds and hundreds of piles of reinforced concrete stabs, rotting, crumbling, rusting from the inside. (…) Weeds had overgrown monuments (…), weeping mildew spread over marble bas-reliefs of a happy worker paradise. (Wendell 2003, 93)

Transition of intention also includes changes in everyday landscapes and practices. Communist-era housing, shopping and cultural facilities have been constantly reinterpreted and confronted with new versions. One of the most important and best examples of the transformational connotations of everyday landscape are the most typical socialist form of mass housing – blocks of flats or *paneláks*. Today they remain home to a mix of social classes, with the middle class prevailing (Stankova 1992). Thus, there is a growing social stigma associated with living in a communist-style block or *panelák*. Many apartments are relatively well located within urban structures and the well-appointed interiors and *panelák* estates, especially in big cities, are the obvious first targets for builders of telecommunication networks, as the housing estates combine a high concentration of people with easy access to underground and in-house spaces for cables. The attitude of *panelák* inhabitants to their buildings vary; some 'people get used to living in *paneláks*' (Stankova 1992, Hanley 1999). Many flats are now the property of their inhabitants and whole buildings are managed by syndicates of flat owners or cooperatives. Many buildings have been renovated, often with the support of local governments. Renovations have often included the installation of thermal insulation for energy efficiency and new coats of colourful paint. Although block of flats apartments were considered highly desirable during the communist era, since 1989 a combination of decreasing population, renovation of older buildings, and construction of modern high standard alternative housing has lead to high vacancy rates, especially in East Germany, or concentrations of low income minorities, like Vietnamese in East Berlin or Gipsies in Czech Republic, Slovakia, Bulgaria and Hungary. At the same time, some conveniently located blocks in central Warsaw, Sofia and Bucharest are still very desirable and expensive.

Transition of intention is the most unbalanced and unpredictable phase of liminal landscape transformation. Many intentions are being consciously and unconsciously redefined. Some blocks of flats look better and better every year and less and less reminiscent of the grey and sad 'machines for living'. Some landscape features have been demolished; others are scheduled for demolition, including blocks of flats in Schwedt, Leipzig, Halle and other towns in East Germany. Other *Large Abandoned Objects*, mainly infrastructural, are being completed 20 or 30 years after beginning of construction, like Gdynia Kwiatkowski Fly-Over in northern Poland, while some other, mainly post-industrial, are transferred into theme parks or end up at the scrap yard. In the next 10 or 15 years, the transitional phase will most likely be over. New layers of interpretation and landscape solutions will be implemented. The 'unbalanced' landscape will either physically disappear or will be reinterpreted and disappear mentally from social memories.

Reincorporation and new construction of old landscapes

Changing contexts might not be enough to transform all aspects of the liminal landscape, but it definitely symbolizes social and cultural transformation. The complexity of the reinterpretation practices is mirrored in various attitudes and contextualization of different social groups. Constructivism in the social sense emphasizes the shared character of language and landscape and relays on the presumption that things don't mean anything on their own; people do construct meanings, using systems of representation, their concepts and signs. Meanings are not conveyed by any feature of the material world, but by the system we are using to represent our concepts. It must be remembered that one object or feature can have different constructivist meanings, dedicated by different social groups. Frequently groups of youngsters use distinct system of representation than an older generation, so places and urban features might have separate constructivist meanings for them. Re-construction is usually related to re-incorporation of de-constructed features and icons. Reincorporation, according to Turner (1975) is the final rite, when the division between 'old' and new' becomes insignificant and eventually disappears or is used in new social roles. That phase might have just begun in Central Europe, and most likely will be implemented by the following generation. Numerous cultural groups create their own systems of representations, based on distinctive construction which results form particular experiences and expectations.

Due to the limited connection with the outside world, the socialist landscape had resisted, to some extent, the globalization flows until the early 1990s. There is a growing demand for grandeur and symbolism in the post-modern world which can be found in many features of socialist cultural landscape. Growing tourist demand and often limited local attractions forced local societies to re-interpret old icons to meet requirements and pressures of competitive markets. The reinterpretation is based on 'museumification', or preserving mainly 1950s features, as symbolic, museum-like objects, forgetting and stripping off the negative meanings. An historical patina makes the pompous landscape and Stalinist heritage quite an attraction, which appeals to many tourists. Some of the grand designs are preserved as architectural and cultural representations of the past times. One of the most popular post-Stalinist urban arrangements includes the 1950s new town of Nowa Huta in Kraków, designed in the neo-renaissance and classicist style known as soc-realism. Similarly, the triumphalist former Stalin Alley in East Berlin, or Poruba district in Ostrava in Czech Republic meets tourist demand for 50-year-old distinctive features (see Chapter 3, New urbanism and new towns; Process of socialist landscape development; Stalinist-era urban landscaping). All of these urban establishments are listed and are present in city sightseeing programmes, as well as guidebooks and tourist maps.[13]

Box 4.6 Karl Marx Avenue, Berlin

The Stalinallee, now Karl-Marx-Allee, is a monumental socialist boulevard built by the young GDR between 1952 and 1961 in the Berlin districts of Friedrichshain and Mitte. The boulevard was named Stalinallee between 1949 and 1961 (previously Große Frankfurter Straße), and was a flagship building project of East Germany's reconstruction programme after World War II. It was designed by the architects Henselmann, Hartmann, Hopp, Leucht, Paulick and Souradny to contain spacious and luxurious apartments for ordinary workers, as well as shops, restaurants, cafés, a hotel and popular cinema *International*. The avenue, which is 89 metres wide and nearly two kilometres long, is lined with monumental eight-storey buildings designed in the so-called *zuckerbäckerstil* ('wedding cake style'), the socialist classicism of the Soviet Bloc. The majestic avenue was completed in 1961 with the modest finishes leaning towards *Jugendstil* and Prussian Neoclassicism. The street would later be extended in International Style idiom and renamed Karl-Marx-Allee. The buildings differ in the revetments of the façades which contain traditional Berlin motifs by Karl Friedrich Schinkel. Most of the buildings are covered by architectural ceramics. Two pairs of domed towers on Frankfurter Tor (see Figure 4.8) and Strausberger Platz designed by Hermann Henselmann are the main landmarks of the Karl-Marx-Allee. On 17 June 1953 the Stalinallee became the focus of a worker uprising which endangered the young state's existence. Builders and construction workers demonstrated against the communist government, leading to a national uprising. The rebellion was quashed with Soviet tanks and troops, resulting in the loss of at least 125 lives. Later the street was used for East Germany's annual May Day parade, featuring thousands of soldiers along with tanks and other military vehicles to showcase the power and the glory of the communist government. The boulevard later found favour with post-modernists, with Philip Johnson describing it as 'true city planning on the grand scale', while Aldo Rossi called it 'Europe's last great street'. The characteristic and renovated tiled buildings are also the scene of TV advertisements, films and music video clips. Since German reunification most of the buildings have been restored and have became a major tourist attraction of the city (Klopeck 2006).

---

13 Czechs seem to be least proud of their socialist heritage. One of the best examples of soc-realism, Poruba complex in Ostrava, is in fact barely present in official promotional materials and practically absent on the city web page. At the same time, Nowa Huta and Stalinallee are fully incorporated in local marketing strategies of Kraków and Berlin.

Figure 4.8 New tourist attractions: Frankfurter Tor and Karl Marx Avenue, Berlin, 2004

A recent trend in Central European countries is the growing process of recognizing 1960s and 1970s modernist design. From some of the participants of post-socialist landscape discourse, modernist structures were the very best of 'their times' and cities: functional, modern, avant-garde, and relevant to the Western stylistic period. Many architects and designers of those buildings are now recognized architectural gurus. Recognition is connected with a new tendency for valuing the recent past.[14] Local and national media cry over plans of destruction of modernist architectural achievements. There is a growing contestation against plans of significant refurbishments of classically brutalist Katowice and Warsaw Central train stations. Some 1970s architecture is highly original or artistic, even though it carries, together with 1950s and 1960s structures, the stigma of being 'wrongly inspired' or carrying 'socialists texts'. Many of the 1970s East German *Zentrum* department stores had been decorated by interesting metal-work façades. In the 1990s most of the centrally located stores were bought by West German retail chains, and the old, but somehow arty, buildings are being replaced by bunker-looking, functional and frequently sandstone-covered new constructions. One of the last remaining, after

---

14 Characteristically, the euphemistic manes of the communist era are quite typical for the Central European 'post-' discourse. Usually when recalling resentments of better past, the post-war period is called 'recent past', while connected with crimes and oppression, it becomes the age 'communism' or 'socialism'.

the demolition of the prestigious Dresden Pragear Strasse shop (in spring 2007) and the innovative and modern structures of Warsaw *Supersam* store, is[15] the 'tin box' in Leipzig (Figure 4.9). The process of public reincorporation discourse is supplemented by the media and various artistic retrospectives and presentations, like 2005's *Unwanted Heritage: Various Faces of the Architectural Modernity* in Gdańsk and Lepzig or Warsaw's 2007 *Concrete heritage: From Corbusier to the Blockers.* New contextualization of functionalist landscape is a difficult task, especially because most late 20th century urbanism generally failed to create considerable landmarks and landscape icons. Additionally, for a substantial part of the Central European societies, hardly functional functionalism and mediocre architecture are deeply connected with the failures of communist planning and landscaping. Negative connotations and associations, together with limited practicality of block design and its social malfunction, leave blocks strongly interpreted as a 'socialist landscape'.

Figure 4.9 Potential modernist heritage: closed 1970s department store, Leipzig, 2007

Old communist-era landscape icons can re-packed and reinterpreted to meet contemporary place marketing demands. A new social context of nationalist pride has been attached to the Civic Centre of Bucharest, and especially to the Palace of the Parliament. Travel guides and brochures proudly concentrate on the magnificence

---

15 Or maybe was. It was still there in summer 2007.

and opulence of the building, constructed only by Romanians, using Romanian raw materials, while the infamous initiator is practically nonexistent in local texts or contexts. One of the most popular examples of re-construction of the cultural codification and re-negotiation is the grand head of Karl Marx in Chemnitz (former Karl-Marx-Stadt), Germany. The huge head stands on a pedestal in front of a tall administration building on the central crossroad of Chemnitz. The three-metre-high pedestal still keeps the distance and almost forces respect. The monument was constructed in 1971 and has been a historical monument since 1994. According to Weiske (2002), 70 per cent of inhabitants questioned see it as a symbol of the city. The bronze head is used in the newest place marketing campaign, promoting Chemnitz as *Stadt mit Köpfchen* (City with the Head). The space around the Head seems to be mainly used by a bunch of local skateboarders, attracted by high-quality granite flooring, and an occasional bemused post-socialist *flâneur*. Interestingly, even though the Head is a formal monument and a symbol of the city, it is hardly presented in promotional materials or postcards.

Another type of social de-construction and incorporation is related to the recycling of old icons and reusing symbols in modern societies and economies. This attitude is generally constructed by the young and the trendy, looking for new inspiration and stimulation. A combination of leftist icons and 1970s design result in quite an attractive product, appealing to many who never experienced communism. Hundreds of items, like copies of old badges, posters, T-shirts and postcards, but also plates and egg cups designed in the communist style,[16] use the old symbols in a brand new cultural context (Brussig 2002). The red star, CCCP[17] or Lenin's head is hardly anything more then an aesthetic sign, trendy and fashionable in some of the social groups. Demand for socialist kitsch, de-sacralized and recycled icons seems to be merely an original and visual trend, sometimes unconsciously promoting forgotten or unrealized ideas, often connected with marketing and meeting growing demand for original, unique cultural products. One of the most popular and imaginative geo-symbols of 'past times' and interesting example of de-contextualization of popular, item which became an icon of East Germany is Ampelmann.

Other new but stylish uses of the old iconic features appear in dozens of post-communist theme pubs and bars, located in many cities around the region. The bars, like Committee in Lublin (see Figure 4.10), PRL[18] in Wrocław or Under Red Hog in Warsaw, are focused on both local clientele and the tourists searching for something familiar and funky. The interior design, is full of communist propaganda and icons, as well as names recalling the communist past, but only in funny, amusing, odd, curious or comical ways. These places are being promoted as 'the last secrets of the Communists', while styled pictures of Marx, Engels, Lenin and other iconic 'saints' complete the interior design.[19] Some of the exhibits are original communist

---

16 Usually made recently in China.

17 In Russian Cyrillic, Union of Soviet Socialist Republics or USSR. Many young people wearing the CCCP T-shirts cannot decode the abbreviation, usually roughly connecting it to 'something leftist'.

18 Polska Rzeczpospolita Ludowa or People's Republic of Poland.

19 Sometimes, like in Wrocław, the communist icons can be also found in toilets.

features, while some others are recent copies. Many of those places are not only full of tourists, but usually also local students, for whom looking for post-socialist past is the way to self-identify in globalizing and amalgamated world. For most young tourists visiting 'socialist theme parks' or pubs, the trip to communist times is as exotic, or often even more so, as travelling to another continent.

Box 4.7 Amplemann, Berlin

The little personalized character, Ampelmann or Street-Light Man, with distinguishing hat, appeared on every red/green street light in East Berlin until 1989. In the early 1990s the sign was considered backward and to be memorizing communist divergence. The municipal proposal of unifying the street lights across Berlin to match 'the better – Western standards' in the mid-1990s met a lot of resistance and opposition from the residents of eastern districts of the city. For many of them fight for the characteristic Ampelmann grew, rather surprisingly, to a symbol of the local identity and 'West colonizing the East'. In 1996 the first product based on the 1961 designed Ampelmann came to the market and soon became the symbol of 'old, better times', as well as an identity symbol of the GDR. The sight slowly came back to many of eastern street and cross roads. The renewed Amplemann became one of the most characteristic, and trade marked, symbols of Berlin, which can be seen not only an street lights, but also on key-rings, mugs, towels, bags and other items.[20] The funky icon embodies the success story of almost gone, but recovered as fashionable and admired symbol (Amplemann 2007).

Figure 4.10 Funky post-socialism: interior of the Committee bar, Lublin, 2006

---

20 Ironically, the present career of the little man in a hat is actually a marketing achievement of West German specialist who came up with the idea, saw a market niche and runs the company.

Memorializing anti-communism

The communist period and associated cultural landscape is being critically contextualized as a time and space of oppression, devastation and tyranny. Meanings and contexts created by *homo instrumentalis* or *homo sovieticus* could only result of oppressive policy, and only remembered as such (Śpiewak 2005). Disgraceful and/or insignificant icons bring the dark memory back, so the 'recent past' and its residua can be merely kept as warning witnesses for future generations, elements of historical education or tribute to the victims of communist totalitarianism. Negative and disapproving constructions of the old communist icons have usually resulted from personal or social memories of repressive actions or connotations. Reminders of grievances, injustices, restrictions and sacrifices are ceded from generation to generation, as an 'anti-communist heritage' and family identity. Deep hurts are very hard to forget, and can be additionally enhanced by long-lasting feelings of revenge and injustice. This pejorative position often denies and rejects any positive developments and achievements of the communist area and principally enables any discussion and compromise. The only reason to reincorporate communist heritage into contemporary memorial policy is to remember the past crimes and as a warning against possible future mistakes.

Remembering anticommunist features is an important element of national historical and landscape policies. Emphasizing resistance, victims of communism and heroism enables to create image of the 'rightness', even in the 'dark ages of mistakes and distortions'. Despite severe suffering and thousands of victims, there are few icons left to remember the dark heritage of the communist period in countries stigmatized by 45 years of totalitarian rule. There are actually relatively few places and memorial landscapes to remember and pay tribute to the fatalities of communist offences. Dozens of crosses symbolize and memorialize victims of communist crimes and can be seen along the former Berlin Wall, as well as on the squares in the city centre of Bucharest, as tribute to casualties of 1989 December revolution. Monuments and commemorative plaques to Soviet and communist sufferers are visible in almost every town and city in Poland, but also in central Prague (see Figure 4.11), Budapest and Sofia. Poland has probably the highest density of anti-communist landscape features, memorializing victims of Soviet aggression in 1939, murdered Polish officers in Katyń in 1940, executions of the Home Army soldiers after 1945, 'hunger strikes', Solidarity trials and many others. Colossal and respected monuments dedicated to the victims of communism have been erected in Gdańsk, Poznań, Gdynia, Katowice, Szczecin, Wrocław and many other cities mainly in the early 1990s.[21] Even a small town like Nowy Staw in northern Poland (population 4,000) has four places commemorating communist crimes. The most recent initiative of the Institute of National Remembrance is to map all the communist crime sites across Poland.

---

21 Gdańsk's Three Crosses monument devoted to the fallen shipyard workers in December 1970 was built in 1980. It was the first sizeable and public memorial of the victims of a communist regime erected in a communist country.

Figure 4.11 Remembering the dark heritage of communism: memorial of the
victims of communism, Prague, 2007

More educative and informative goals are being realized by museums and centres of anti-communist reminiscences. Memorializing victims of communism and anti-regime resistance became a part of important political projects, which were instrumental in establishing some of the region's leading anti-communist documentation centres and museums, including Gdańsk Solidarity Museum in former Lenin Shipyard, the Berlin Wall Museum, Forum of the Historical Times in Leipzig, Bucharest Museum and Park of Totalitarianism and Socialist Realism and Prague's Museum of Communism. There are some museums commemorating communist crimes located in former state security quarters, like Budapest's House of Terror (Figure 4.12) and Berlin's Stasimuseum, and in ex-prisons, like Jilava Memorial Museum in Romania. Some of those museums, like the one in Prague or Budapest, are private initiatives documenting evidences of crimes of communist legacy, while the role of central governments, excepting Germany, is usually limited to statements and celebrations. The situation is not different even in often loudly anti-communist Poland, where anti-communist heroes and events are parts of the official historiography, but there is no museum dedicated to the 45 years of socialist history of Poland.

Figure 4.12 Landscape of terror: House of Terror, Budapest, 2005

Box 4.8 Museum of Communism, Prague

In 2001 the Museum of Communism was opened on central Prague's fashionable Na Příkope Street in rooms sub-leased from the ultimate symbol of capitalism, McDonald's, who have an outlet downstairs in the same building. The Museum of Communism was established by a young American businessman, owner of the Bohemia Bagels shops. Visitors get to see everything from the statues to a mock-up of a communist torture room. The Museum presents a vivid account of communism, focusing on Czechoslovakia in a variety of areas, such as daily life, politics, history, sport, economics, education, arts, such as the so-called Socialist Realism movement, media propaganda, the Peoples' Militias, the army, the police, censorship, judiciary and coercive institutions. The museum focuses on the totalitarian regime from the February coup in 1948 to its rapid collapse in November 1989. The original items and meticulous installations containing authentic artefacts are displayed in the three main sections: the Dream, the Reality, and the Nightmare. The filmmaker and exhibition curator, Jan Kaplan, who escaped his homeland and fled to London during the Prague Spring of 1968, describes the museum as 'a tragedy in three acts'. The first features propaganda material and a classroom with communist school books and is meant to portray the idealism some people felt in the early days. The next part is the nightmare – the harsh reality of communism, everything from empty shop shelves, to video clips and an interrogation room. The last section focuses on the 1989 Velvet Revolution. Highlights from the displays include rare items from the Museum's own comprehensive archive as well as material obtained by the organizers from major collections, both public and private. The Museum of Communism, it is declared on the Museum's web page, was created for the display and interpretation of objects and historic documents and it stands as an authoritative historical narrative relating to this 20th century phenomenon (Museum of Communism 2007).

Landscape icons were often left as a warning and intentional realization of particular anti-communist historical policy. Tributes to the heroes of anti-communist resistance were often more then just memorials to victims, but have been used for contemporary political purposes by conservative and right-wing parties. The new memorials are part of an ongoing process of commemoration rather than being static objects that, once erected, are gradually forgotten. Their construction has drawn communities into debate about the past, and, once dedicated, they serve as the focus for new rituals of commemoration (Fote, Tóth and Arvay 2000). 'When justice does not succeed in being a form of memory, memory itself can be a form of justice', said Ana Blandiana, the founder of the International Center for Studies on Communism, based in Bucharest (Memorial of the Victims of Communism 2007). The 'reminding to remember' policy reflects attitudes of local and national authorities, but also many ordinary people.

In March 2006, Marius Oprea, the President of the Institute for the Investigation of Communist Crimes in Romania, noticed that 'many people are asking whether an outright condemnation of communism is still necessary, now that 16 years have passed since the Revolution of 1989. I say it is, and I rely on the fact that such a profoundly moral decision, meant to restore the Romanian society from its foundation, is necessary at any time.' Moreover, the setting up of the Institute for the Investigation of Communist Crimes in Romania[22] took place at a time when civil society was accusatory of the lack of decisive reparatory actions on the part of the public institutions which were assigned to deal with communist crimes and abuses. In addition, restricted access to archives fed the idea that it was a lack of political will which hampered the reestablishment of truth and justice (Institute for the Investigation of Communist Crimes in Romania 2006). New construction and elimination of unwanted and communist associated meanings can be enhanced by existing or prepared legal acts, like in Hungary and Poland. The tendency to recap and accentuate the negative, criminal aspects of communist times is connected with activities of anti-communist, right-wing parties, like Law and Justice (*PiS*) in Poland. Remembering misfortunes, traumas and victims is an important part of national and local identity, but can become also a political tool, aimed at achieving short-term goals.

Box 4.9 Memorial of the Victims of Communism, Sighet

The prison in Sighet was built in 1897, as an ordinary law prison. After 1945, the repatriation of former prisoners and deported persons from the Soviet Union was done through Sighet. In August 1948 it became a place of imprisonment for a group of students and peasants. In May 1950 over 150 dignitaries from the whole country were brought to the Sighet penitentiary, including former ministers, academics, military officers, journalists, bishops and politicians; some of them were convicted and sentenced to heavy punishments, whilst others were not even judged. The penitentiary was considered a 'special work unit', known as 'Danube colony', but in reality was a place of extermination for the country's elites and at the same time a place with no possible escape, with the frontier of the Soviet Union being less than two kilometres away. The prisoners were kept in unwholesome conditions, miserably fed, and stopped from lying down during the day on the beds in the unheated cells. Humility and ridicule were part of the extermination programme. In 1955, following the Geneva Convention and the admission of communist Romania to the UN, some pardons were granted. The prison once again became an ordinary law one. However, political prisoners continued to appear in the following years, and many were kept secretly in the local psychiatric hospital. In 1977 the prison was put out of use and entered a process of degradation. The Civic Academy Foundation, established by Ana Blandiana along with 175 others, took over the ruin of the former prison in 1995, having as its general purpose civic education and as its immediate objective the creation of the Memorial. Each cell became a museum room, where objects, photos and documents were placed, creating the environment and documentation of a museum. In one of the internal yards of the former prison, following a competition of projects in which 50 architects and artists participated, in 1997 a Space of Meditation and Prayer was built. On the walls were engraved the names of almost 8,000 people who died in prisons, camps and deportation places from Romania. In the early 2000s, a conference hall was added together with a number of objects of art including the tapestry 'Freedom, we love you'; the painting 'Resurrection'; the sculptures 'Homage to the political prisoner'; the statuary group 'The convoy of martyrs', which has become one of the emblems of the museum; and the Cemetery of the Poor, situated 2.5 kilometres outside the city (Memorial of the Victims of Communism 2007).

---

22 Acknowledged by the Romanian Government's Decree No. 1724/ 21.12.2005.

Less benign symbols of the communist years are gaining new memorials. These include the prisons and forced-labour camps in which the communists incarcerated and punished their opponents. Many of these prisons are still in use, so plaques and memorials have been affixed to the exteriors, but some are transferred to a memorial site, often financed by a private foundation (Fote, Tóth and Arvay 2000).

Sites and memorials of anti-communist riots and revolts became important places of public manifestation in Poznań, Gdańsk, Gdynia and Szczecin in Poland, as well as in Prague, Bucharest, Leipzig and Budapest. The squares and streets that witnessed mass riots and protests during communist times, like Wenceslas Sq. in Prague, St. Nicolas Sq. in Leipzig or Mickiewicz Sq. in Poznań are still among the most significant civic spaces in the cities, while their meaning is in enhanced by monuments and memories of anti-communist struggles. Ironically, comparing to all the other iconic places, the least commemorated seems to be the first and one of the largest, the 17 June 1953 Berlin riots, which might somehow represent East German reserve over negative commemorations of their communist past.

Probably the most important and best-presented part of the anti-communist landscape is dedicated to remembering the Autumn of Nation of 1989. Icons of the Fall have been caring, positive and victorious messages; are also most recent and best remembered, not only in official, but also in private memories. The collapse of the Central European communist regimes is memorialized by numerous Polish Solidarity or *Solidarność* trade union trails and places. From summer 1980 until spring 1989 numerous factories' strikes, street fights, churches, 'interning camps' and sites of the communist crimes and political murders have became places considered to be milestones on the 'roads to Freedom'.[23] Mass gatherings and silent protests in Leipzig's St. Nicolas church are evoked by a palm-shaped column and commemorative plaque. Protests in Romanian Timişoara and the bloody December Revolution in Bucharest are commemorated by memorials, crosses, shrines and monuments in many places in the Romanian capital, while some of the bullet holes have been left on the façade of the University's Humanities building on the Bucharest Revolution Square. For many younger members, post-socialist heroes, tortures and restrictions are as distant as the Napoleonic Wars, and as often much considered. Kveta Libenska, imprisoned by the Czechoslovakian authorities during the 1950s, said 'young people don't know what things were like, and I doubt they will ever understand what we went through' (see Museum of Communism 2007).

Ostalgic landscapes and representations

The turn of the 1980s and 1990s was marked by great changes in Central and Eastern Europe, simultaneously stimulating expectations of a new and better future for the newly liberated nations, which sooner or later initiated processes of democratic transformations. From the perspective of nearly 20 years, it has become evident that these processes will be continuing for many years to come. The most telling

---

23 'Roads to Freedom' is also a title of Poland's largest Solidarity memorial place and exhibition located in the former Lenin Shipyard in Gdańsk, known as 'the Solidarity cradle'.

proof of it is the phenomenon of nostalgia for the 'recent past' that casts a shadow on the political reality, condition of societies, and sometimes also on the culture of the period of transformation. A vision of the future must account not only for such unresolved problems as unemployment, ever-wider differences in social status, developmental barriers to a market economy or ethnic and nationalistic conflicts, but also for the stubbornly returning sentiment for the old times. The past has become more topical than before: it is a burden and a problem, a subject of content and dispute, as well as – and this is intriguing – a source of longing (Modrzejewski and Sznajdeman 2002).

Post-communist melancholy is expressed in culture, social behaviour and political choices. Awareness that express a longing for communism – perceived not as a political system, but as a closed era which in popular opinion symbolized a now-lost sense of security, and also employment, equality, cult music, films and books. The enthusiasm of the first years of regained freedom soon turned into disillusionment, while the ever-wider phenomenon of post-communist melancholy shows that transformations have elicited a sense of insecurity and anxiety. The loss of a symbolic centre, the downfall of opinion-forming bodies, a diminishing trust in politicians and the waning role of milieux which previously had masterminded the transformation strategy, the growing problem of expendable people and widening differences in social status – all those factors have led to unexpected changes in what people think of themselves in social, cultural and political terms. Beliefs and ideals of the past have fallen into public oblivion, while faith in a spiritual revival and the legend of social solidarity have given way to deeply pessimistic views of the social condition. The loss of trust in politicians and public institutions as well as a sense of exclusion have brought about the defensive mechanism of nostalgia – not only a longing for the safety of childhood and youth, but also the projection of personal memories and emotions on historical opinions and participation in the collective process of obliterating the past. In other words, we are confronted with the creation of a new collective identity which is real only in the collective imagination (Modrzejewski and Sznajdeman 2002).

There is a growing realization that any kind of totalitarian regime, Nazi, apartheid or communist, did not only produce resistance. Rather, alongside difference and inequality lie more subtle forms of economic, cultural, and intellectual exchange integrally tied to the layers in which past and present are negotiated through memory, tradition and history (see Minkey and Rassool 1998). Remembering the socialist past has been as selective as any process of forgetfulness and oscillates between carnival, museum, golden times of youth and the promised 'workers' paradise'. Sometimes the emotional attitude towards the post-socialist landscape mirrors the nostalgic sentiment of the older generation. The socialist landscape sometimes reflects and resembles 'the old good times', stability and missed youth. Nowhere is the process more obvious than in East Germany, where the *Ostalgie*[24] stands for the longing of the German Democratic Republic and, as it is remembered by many, better and happier times (Brussig 2002). Similar yearning for old, better times can be recognized in substantial aspects of the post-communist parties' electorate all around

---

24 The word *Ostalgie* is merges German *Ost* (East) and *Nostalgie* (nostalgia).

the region. There are similar mechanisms of conservative nostalgia of socialist social order and post-communist nostalgia of a system, although nostalgia should not be mistaken with memory. Memory brings the past how it was – with its good and bad aspects, while nostalgia tends to idealize and focus only on positive connotations, or a part of the truth. Communist nostalgia is connected only to the last, mainly in the 1970 and the 1980s period of the communist history; not many want to remember and recall the 'dark ages' of the 1950s and only a slightly better 1960s. As Esterházy (2007, 14) puts it, 'this era is petrified with us, we can read it in our fits, elaborated in that time. And this happens despite the fact, that consciously we might not want this memory to come back.' It is still generally the yearning for comfort and security, mixed with the longing for youth, but it is hardly nostalgia for the communist system per se (see Šimečka 2002).

Every museum, especially historical museums, are always political projects, where the organizers try to stress and commemorate some aspects of the past, since it is never possible to present all of the features of the historical discourse. Museums are also representation of ideas, beliefs, thoughts and maybe fears of their creators. There are some nostalgic exhibitions and museums in East Germany, including the newest GDR Museum in Berlin. Some other museums and exhibition centres commemorating rather positive aspects of the communist era carry rather peculiar names, like *Zeitreise* (Travel in Time) in Radebül near Dresden or very informative, like *Dokumentationszentrum Alltagskultur der DDR* (Documentation Centre of Everyday Culture of the GDR) in Eisenhüttenstadt. The names do not directly reference the communist period, as if trying not to raise too much controversy and antagonism or too much negative remembrance.

Box 4.10 DDR Museum, Berlin

The nostalgic feelings and longings can be probably best exemplified in private GDR (DDR) Museum in Berlin (Figure 4.13), opened in central Berlin at the meaningful address Karl-Liebknecht Street 1 in 2006. The museum was established by a group of East German businessmen collecting original memorabilia. DDR museum is small, compact, but holistic, as well as interactive. It looks like a bit of time capsule of the late Honeckerian, happy everyday life. The DDR Museum examines life in the German Democratic Republic 'from every angle: you can watch TV in an authentic GDR living room, rummage through the drawers of the Carat wall unit, have a sniff at the spice rack in the kitchen and marvel at the pressure cooker left on the stove'. The project might appeal both to *Ostalgic* seniors and trendy tourists, since 'the spirit of an epoch is not just reflected in pictures and books, but also in pots and frying pans'. The Museum promises to 'experience daily life as it was' but only for some, only sometimes and only in some places. As it is declared at the official web page, 'whether it concerns shopping habits, mobility, recreational activities, work, family or home décor, the GDR Museum Berlin offers great diversity – objective in its facts, subjective in its view of the situation'. The subjectivity of the discourse is expressed in general absence of the negative aspects of the East German regime; only the Wall is presented, as an introduction to the exhibition (DDR Museum 2007).

Figure 4.13 *Ostalgic* representation: interior of the DDR Museum, 2006

In recent years various facets of the ever-living nostalgia has been increasingly affecting politics, culture, artistic life, results of public opinion polls, social awareness and the sphere of ideas and values. Stories about homelands that sometimes do not exist any longer in many ways tackle the issue of remembrance and what is remembered. Personal experience is often combined with an idea, a symbol with the reality, the history and one's intimate biography. In this way, in such a multitude of perspectives and voices, one may find common features of those nostalgic images and longings, processes affecting the collective memory of societies in post-communist countries and attempts to comprehend them. They form a picture – necessarily selective and subjective – of the contemporary world of our identity at the time of transformation, but there is a danger behind *Ostalgia*. *Ostalgic* re-interpretation of the socialist past and its icons might imply falsely positive and embellished image of the times. One old man interviewed said that 'after German unification it seems, that all our life In GDR had no sense and no significance, that everything was bad'. Many symbols and features have been deliberately eliminated and replaced by better, Western goods, so for some, appreciation of the *Ostalgic* landscape can be related to appreciation of his or her life and achievements (Ampelmann 2007). Reminding the past and one's youth is natural, but always subjective. Remembering the 'old, good times'; buying fancy memorabilia, like plastic hen-shaped GDR egg-cups; recalling stability and 'social justice'; must be accompanied by the realization that all the communist regimes survived 40 years only due to ruthless dictatorships, terror and surveillance.

Our relationship to the socialist past is a mixture of the memories of generally trouble-free youth, some sentiment and some relief. Sentiment of past decades of communist regimes brings the memories of the younger years, when we all were 'slimmer, healthier and more beautiful', and the social state offered stable jobs, free health care and holidays. At the same time we rather forget the miserable salaries, aggressive and oppressive militias, political prisons, low-quality consumer goods,[25] censorship and travel limitations. Devoid of negative memories and meanings, the happier picture and text of communist past is being transferred to the younger generations through family stories, films and gadgets. Students buy only the part of the communist heritage that has been incorporated into the mass culture. Communist icons, like Che Guevara, have been incorporated into pop culture devoid of any negative historical or moral connotations. The hoarse aesthetics of the Eastern Bloc design seem to be an alternative to mass products of H&M and Ikea because they are highly original and distinct. Selective memories are not corrected by serious historical discourse, so the past is to some extent remembered through funky icons and texts. Cultural icons like cars (Trabants, Skodas, Dacias, Polish Fiats), household goods, designs, emblems and films, transfer us to the country and landscapes of youth we remember or to exotic historical places (Jaroszyńska and Jędrzejczak 2007).

---

25 And in some countries, especially Poland, low or close to no quantities.

Chapter 5

New Landscape Symbols of 'New Europe'

The transformation of post-socialist cities is facilitated not only by anti-communist reactions, but also by a set of contemporary social, economic and cultural processes, which shape the urban landscape of Central European cities. After socialism, cities faced vast legal, structural, cultural and visual conversions. The changes have been accelerated by the explosion of the free market and flow of capital, reintroduction of land rent and privatization, as well as the appearance of new actors on the landscape scene, including local governments, free media, private owners and investors, as well as inhabitants and NGOs. Accumulation of need, capital and power has been manifested dramatically in urban settings. Post-socialist landscape transformation is most clearly visible in large cities and metropolises. Many smaller towns and villages are generally characterized by more stable scenery, slowly reacting to major but distant events and powers.

An important part of the post-socialist transition is the re-naming and re-branding of the region. The notion of post-communism always carries, at lest for the local population, a notion of an unwelcome communist legacy. Since a 2003 comment by Donald Rumsfeld who, when asked about the European opposition to the Iraq war, made a division between 'old' and 'new' Europe, the phrase 'New Europe' has became famous. Although 'New European' countries were originally distinguished by their governments' support of the 2003 war in Iraq, as opposed to an 'Old Europe' unsupportive of that war,[1] the term is more broadly used today for everything from film festivals and BBC travel series to financial markets.[2] The rhetorical idiom 'New Europe' was used by conservative political analysts in the United States to describe European post-communist-era countries, but soon after the Rumsfeld term was widely adopted in the region, and many post-communist countries quickly adopted it and declared themselves to be the 'New Europe', to much acclaim for their courage and integrity. Now 'New Europe' is used to describe the rising power, prosperity, and quality of life of those Central European countries formerly associated with the Soviet Bloc. This is in contrast to those European nations that are historically better

---

1 The usage of 'New Europe' in reference to its position on the invasion of Iraq has been criticized (Chomsky 2004). Even initially the usage of the term was not based solely on this fact. The governments of several other countries, such as the UK, Denmark, the Netherlands, Italy, Portugal and Spain also supported the war, but are not commonly considered as belonging to a 'New Europe'.

2 In October 2007 the *New York Times* website had more then 500 references to 'New Europe', understood as post-communist Central Europe.

known to the West, but whose pre-eminence is perceived to have been in a relative decline since the early 1990s (see Shanker and Mazzetti 2007, Chomsky 2004).

Post-socialist resolutions of the early 1990s were characterized by a fairly spontaneous understanding of freedom on both personal and institutional levels. After more than 40 years of oppression and restraint, control mechanisms virtually disappeared. Party leaders, apparatchiks and industrial 'red barons' vanished from the landscape management scene, as well as from the omnipotent central planning bureaux. Procedure of landscape supremacy had been seceded to local self-governing bodies, frequently somewhat unprepared for the new challenges and responsibilities (Czepczyński 2005a). A strong belief in personal freedom was quite often adopted and became a significant landscape management canon. A typical American-style landscape blend, with modern glass and steel constructions dominating vertically, surrounded by a dilapidated setting, is one of the most typical post-socialist landscapes, and is a result of a specific, even if unintentional, landscape management practice. Radical changes in urbanized landscape administration resulted in spatial confusion and a certain level of anarchy. Newly elected self-governments, both at regional and local levels, had to cope with repeatedly changing regulations, as well as high expectations from the local communities. Recently freed inhabitants, released from the ruthless chains of socialist regulation, were expecting to enjoy the rights of private ownership, to an extent seldom met in Western European countries. There was a popular opinion that 'my land is my castle, and now only I decide what my landscape will look like'. Personal taste, as well as national heritage and financial recourses, were mirrored in the features of the urban landscape of the early 1990s.

The general outlook of our cities has been radically changed and is reflected in designs, everyday practices, recourses, possibilities and aspirations. One of the first, quite minor, but very perceptible and visual rehearsals of post-socialist cities was the massive infusion of lights, colours and noise (see Bauer and Wicker 2004). The socialist economy was not too keen on wasting strategic resources, and almost everything was strategic in communist times, including paint and illumination. Sad, rather miserable grey and dark after dawn cities of the 1980s have been steadily changed by supplemented brightness, glow and fresh colours. After living in colourless countries with tired people, societies have been overrun by colours, sounds and movements (Czepczyński 2005a). In a few years time, even without any major or structural investments, the aesthetic appearance of practically every post-socialist country had been brightened up by much more colourful and 'Western-looking' crowds and cars, refurbished shop windows, infusion of grand scale and mass advertisements, accompanied by loud music and traffic noise (see Bauer and Wicker 2004).

The new aspects of the cultural landscape of post-socialist cities are negotiated between new economic and political powers and visualized in new developments, new hierarchies and new exclusions. The exclusions and hierarchies are constructed in space and landscape, by means of property rights and rents, laws, transportation systems, and other, more symbolic forms of control. Dominant economic and political institutions carve their imprint on the urban landscape by producing what Lefebvre calls 'abstract space'. This space is delineated and defined by capital investment and prestigious public and private projects, but at the same time these projects brutalize and dominate the city (Zukin 1993). New landscape symbols have appeared in every Central European city to accompany economic and social makeovers and express the new hierarchies, needs and expectations of 'renewed' societies.

Functional transformation of post-socialist landscapes

Since the 1990s modern global trends have affected every post-socialist country. The transformation has been additionally enhanced by global economic challenges, including de-industrialization, globalization and neo-liberalism. Global players and rules, together with the implementation of neo-liberal policy, quite rapidly entered post-socialist economies and socialites. The flow of capital, thought and culture was hardly limited and controlled, and noticeably and severely changed local and national orders and systems. Liberal parties introduced liberal economies, while global markets created new modes of communication and new lifestyles. Non-interventionist landscape management models and tools were often considered to be the very best solutions to post-socialist administration procedures and were widely applied in many post-socialist countries and cities. The openness to global economies brought many global players, including growing financial institutions, real estate companies, international hotels and large-scale advertisers, as well as shopping malls and supermarkets which symbolized the desired Western lifestyle. In some cases, the open and willing nature of societies tired of socialism welcomed those features more enthusiastically than in Western Europe; thus post-socialist landscapes can sometimes seem even more Westernized, or rather Americanized, than many Western European cities (see Sýkora 2007).

Implementation of neo-liberal projects had been recognized as the only and perfect solution to any economic turbulence, resistance or contra-action. In many post-communist countries, like Poland and Hungary, but also Czech Republic, Slovakia and Romania, after the initial triumph of the neo-liberal 'new right', came the renaissance of the 'old right', critical of the mistakes of free market (Gray 2007). Neo-liberal politicians and economists believe that the free market is a crucial and basic condition of personal freedom: even democracy should be limited to protect market rights. The free market is the most efficient economic system and that why it has been spontaneously developing in all regions of the world; while expansion of the free market reduces sources of potential conflicts (see von Hayek 1988). The neo-liberal paradigm of privatization and commercialization of everything that can be privatized or sold seems to be popular in practically every post-socialist country, and has been realized by conservative, liberal, as well as social-democratic governments. This form of general privatization is simply the easiest solution, both technically and intellectually (Nawratek 2005).

Post-communist economic transition was abrupt and aimed at creating fully capitalist economies in the shortest possible time. All the countries of the region had immediately abandoned the tools of communist economic control, especially omnipotent central planning, and moved more or less successfully toward free market systems. The consequence of the rapid and deep systematic transformations was that many of the communist-era grand industries and enterprises collapsed, as they were totally unprepared economically and socially for free market rules. In economics, 'shock therapy' refers to the sudden release of price and currency controls, withdrawal of state subsidies, and immediate trade liberalization within a country. Many emerging economies of the late during 1990s applied 'shock therapy', promoted by prominent economist Jeffrey Sachs. Poland is often cited as an example of the use of shock therapy. In 1989 the government took Sachs's advice and immediately withdrew regulations, price controls and subsidies to state-owned industries. Many economic factors were immediately applied, but privatization of state-owned enterprises was delayed until society could safely handle the divestiture. The policies of shock therapy provoked much debate, which centred on whether or not the final achievements justified the pain that accompanied such radical restructuring (see Paci, Sasin and Verbeek 2004). Sometimes the economic 'shock therapy' produced rather 'shocking landscapes' (see Figure 5.1), representing liberalization of economic and design constrains. Liberated or emancipated landscape lords often applied 'mix and match' strategies to answer shockingly fast developments, especially during the early 1990s. These landscape projects were most often half successful: the functional, significant and structural 'mix' worked quite well, while the 'match' was very seldom achieved. In result many operative, visual and evocative 'mish-mash' landscapes enriched urban discourse of Central Europe (see Wegenaar and Dings 2004, Hauben 2004).

Figure 5.1 Shocking landscapes: shopping centre in western Budapest, 2005

The contemporary city is visualized as a capitalist accumulation of cultural signs and symbols. Every cultural meaning is becoming a commercial artefact. Branded so as to appear necessary to the consumer, the success of these commercial objects is supported by the desire to be a part of or to hold ownership of them. Culture is a powerful, if not the most powerful means to control the city. The *symbolic economy*, as we can call all this management of cultural consumption, puts aside what does not work, meaning that it refuses anything unsuccessful (Zukin 2000). The city, its landscapes, buildings, features and meanings can be compared to stock market activity, where stock is only attractive as long as it sells and as long as the company produces revenues for its shareholders. In the same way, cities and private corporations are forced to produce their symbols in order to survive within the globalized capitalist market. They create their tourist and consumption industries by producing both symbols and spaces. Urban landscape becomes entirely *ageographic*: each of the centres and places is so autonomous that it could easily be emplaced anywhere else (Cupers and Miessen 2002).

The reforms generally helped to increase the common standard of living, but also introduced many negative aspects to everyday life and landscapes that were absent in socialist cities, like unemployment and vast economic stratification of society. For some, like Barber (2007), the tyranny of the state is being replaced by a new tyranny of the free market. This tyranny often looks like freedom, because it allows choices, however limited, but this despotic marker frequently deprives societies of possibility of common acting for the common good. The cultural landscape is being transformed according to new, global expansions of means and money. New economic impulses, however generated, are as likely to require new building, much of it in the central historic districts, as to provide new uses for old structures. Historic Bratislava has been 'invaded' by new developments for the burgeoning financial sector, and new real estate investments in Leipzig have stressed new building rather than rehabilitation (Ashworth and Tunbridge 1999).

Outbursts of unrestrained consumptionism were among the most characteristic aspects of urban emancipated landscapes. After decades of economies of shortages, Central European societies heartily welcomed new possibilities of supply. Markets, and especially the promptly developed hypermarkets, were among the first bearers and providers of the new, affordable, promised life. They symbolized to some extent freedom and welfare, so local societies welcomed them much more enthusiastically then anywhere in the West, at least in the 1990s. Central European shopping facilities have undergone several alterations since 1989. Many of the communist-era state or co-operative shops have become derelict or sold; only some shopping co-ops are still operating in local markets. Popular in the early 1990s, street sellers and *suitcase mobile shops* were, in the mid-1990s, replaced by mainly foreign and peripherally located hypermarkets. Since the late 1990s smaller, often foreign, discount stores grew in central and residential locations in the cities and towns around the region. At the same time, large, centrally located shopping malls were constructed in major cities, introducing global trademarks and global 'sameness' into the region (Ritzer 2007). The ongoing privatization of public space is creating another 'brave new world', devoid of any of the negative aspects of urban life. Sheltered from the possible dangers of everyday life, shopping has become one of the most important recreational activities. Secured by high-tech private security companies and controlled by closed-circuit television surveillance, it could be referred to as the realization of a middle-class-utopia. To quote Jean Baudrillard, this development has been described as the appearance of a new, media-guided society 'increasingly regulated by absorbing *simulacra*: exact copies of originals that no longer exist, or perhaps never existed in the first place' (Soja 2001, 101). But the hypermarkets and huge shopping malls have also been symbols of an emerging middle class and their aspirations. For years Warsaw has competed with Budapest in developing Central Europe's largest shopping centres. In 2007 the number of shops in the newly completed Złote Tarasy (see Figure 5.2) or Golden Terraces in Warsaw outnumbered the Budapest West Centre, but probably not for long (see Gądecki 2005, Neelen and Dzokic 2004).

Figure 5.2 Simulacrum under glass roof: Złote Tarasy mall, Warsaw, 2007

Post-communism faced great demand for modern office facilities, together with many other functional buildings, hardly present or undermined during the communist regime, like high-quality hotels and financial services. Many development projects had been realized to meet that growing demand, accelerated by the rapid economic growth of the region. High-quality office complexes have been constructed in every major city of the region, but probably the highest growth and most spectacular is the Warsaw central business district development. Numerous office towers and hotels were built in the western part of the Polish capital, giving a much-appreciated

'Manhattan look' to the city. Foreign investors played a very important role in facilitating post-socialist landscapes. In the 1990s, only foreign capital could finance sizable developments. In addition, American and Western European architects were employed by Western companies. Since the late 1990s some of the world's most eminent architects have designed prestigious offices in Prague, Warsaw, Budapest and other major cities. New money had to be properly exposed and promoted, while the 'star names' could create remarkable framing and sometimes, like Getty's 'Dancing House' or Fred and Ginger in Prague (see Figure 5.3) on Rašínovo nábřeží, which has become a tourist attraction and new symbol of the city.

Box 5.1 Arkadia Shopping Centre, Warsaw

Arkadia is one of Central Europe's largest shopping centres, boasting the stores selling domestic and foreign brands, restaurants, cafés and a cinema, all of them under a single roof, surrounded by landscape of *simulacrum*: glass rooftops, high street-like galleries, mosaic tiling and natural stone walls. The developers believe it is a 'favourite meeting place of stage artists, singers, media people and foreigners', as well as 'a stylish, elegant, friendly and fashionable place'. Motorized customers may use 4,500 parking places in underground garages. The 287,000-square-metre-space houses almost 230 outlet points, including 184 stores, 15 service outlets and 30 restaurants/cafés. Each level of Arkadia has a specific key profile: the ground floor is primarily intended for everyday shopping with a hypermarket, sports stores, home supply stores with household appliances, multimedia and decorations. In addition, five restaurants are located on the ground floor with separate external entrance doors at the front of the building, so that they are accessible irrespective of the Centre's business hours. The first floor is dedicated to fashion and accessories and houses mainly clothing stores of the Arkadia's 'World of Fashion'. The second floor is for leisure, with a Cinema City multiplex of 15 screens. As the official prospectus says, 'the concept behind *Arkadia* was to rediscover the essence of trade and its importance for the growth of the city'. The design of the Centre was inspired by the city. Its thematic malls carry locally anchored names, like Copernicus, Vistula, Canaletto and Twardowski and were designed to evoke the concept of 19th century European galleries. The public space in front of Arkadia, which includes a city square with a wide choice of catering establishments and a park, is supposed to be 'an excellent place for leisure, a walk, sports and cultural events'. The layout and names of the malls 'evoke diverse areas of human life, thought and environment: the science, culture, myths and the city itself' (Arkadia 2007). The mall, marketed as 'life in its prime', became for some of the Warsawians one of numerous Arcadias, an idyllic location or paradise, where the consumer's dreams come true. But every Arcadia is also a Utopian project, ironically and contrary to the Greek Arcadia; Warsaw's Arkadia is completely created and corrupted by the civilization, where the dreams are for sale only.

Figure 5.3 Trendy new icons: Ginger and Fred buildings, Prague, 2004

By creating self-assured images and symbols, private companies try to stamp an artificial collective identity onto the consumer. Apart from the production of symbols and landscapes, there is the actual production of spaces as a way to control the city and its consuming public. The mental representation of the city that is being provided adds to the soap-opera-like, visually selective, culture of consumerism, with its main goal of raising profit. These places are characterized by strategies to enhance the visual appeal of the city, and thus the rate of consumption. 'The urban policy in the age of consumerism reflects the clear opinion what should and what should not be displayed into city's social matrix. As a visual result, the city has progressively been transformed into an accumulation of well-designed spaceships' (Cupers and Miessen 2002, 23). Many local governments in Central Europe tried very hard to attract symbols of new consumptionism. One of the first visible aspects of post-socialist landscape change was the wide spread of McDonald's restaurants and other mostly American fast-food facilities, like Pizza Hut and so on. These ventures symbolized the desired Western lifestyle, and were very clearly visible in the landscape of many urban centres of Central European cities. For some municipalities and local societies the location of McDonald's have been considered as a sign of prestige and development. The long queues waiting patiently in front of the first McDonald's undoubtedly marked the rank and role of the features in local urban landscape.

Since the fall of communism in 1989, several communist-era buildings have been refurbished, modernized and used for other purposes. Perhaps the best example of this is the conversion of Bucharest agro-alimentary complexes into shopping malls and commercial centres. These giant circular halls, which were most often known as 'hunger circuses' due to the food shortages experienced in the 1980s, were constructed during the Ceauşescu era to act as markets, although most were left unfinished at the time of the Revolution. Modern shopping malls like Bucharest Mall, Plaza Romania and City Mall emerged on pre-existent structures of former 'hunger circuses'. Another example is the modernization and conversion of a large utilitarian construction in *Centrul Civic* into a Marriott Hotel. This process of adaptation and functional transformation was accelerated after 2000, when the city underwent a property boom, and many communist-era buildings in the city centre became prime real estate due to their location. In recent years, many communist-era old apartment blocks have also been refurbished to improve the city's urban appearance (see Ioan 2007, Centrul Civic 2007). Functional alterations were hastened by de-industrialization. Several old production sites, often centrally located, are being redeveloped and re-contextualized into shopping and office facilities and, most recently, into very popular loft accommodation. Post-industrial developments include Faktoria in Łódź, Poland, an old textile factory refurbished into modern shopping mall, hotel and cinema complex and the Old Brewery in Poznań re-assigned to a retail and cultural centre.

Figure 5.4 New businesses in new places: Prague Business Park, 2007

New functional elements have been implemented into the 'New European' landscapes, as copies of mainly American solutions. Numerous business and research parks, together with greenfield industrial developments, have changed the landscapes of most of the regional metropolises. Many office buildings are grouped together, often on suburban locations, where it is cheaper to develop land because of the lower land costs. They are also often located near motorways or main roads. Criticism of business parks often relates to the failure of business parks to relate to the urban fabric of the city. Despite some reserve, the parks are a very popular form of urban development, seen in every country of the region. Establishments like Kraków Business Park in Zabierzów, Business Park Sofia, Orco Business Park in Budaörs, West of Budapest, Business Park Rudná I in Prague (see Figure 5.4), Bucharest Business Park or Warsaw's Mokotów Business Park look very much the same, as does almost every modern office complex.

Cars were considered symbols of personal freedom of movement in communist times. Quite suddenly in the 1990s almost everybody could afford a car.[3] The new wave of massive motorization met our roads and cities unprepared. A lack of parking lots caused parking chaos. At the same time, cars created new possibilities for tourism, entrepreneurship and urban sprawl. The rapid development of motorization and an increase in the number of cars was not followed by the development of public transportation, road and parking facilities. In few years after 1989, many metropolises of the region faced severe traffic problems, similar to, and often much worse than, that of many Western cities. Almost all public spaces in urban centres have been occupied by cars, and many cities, including Warsaw, introduced some kind of parking regulation as late as the early 2000s. Cars for many new owners, especially men, developed into almost cult objects, while some of the car parks looked like temples of this 'new religion' (see Figure 5.5).

Pre-accession and structural funds have been additionally transforming landscapes of post-socialist cities. Massive road construction projects, together with many environmental investments, have been changing landscape of Central Europe since the early 1990s. Piles of sand and construction machinery, together with route diversions, accompany society and tempt its patience. At the same time our rivers and lakes are considerably cleaner, roads and sidewalks less bumpy. Construction sites are probably the most iconic, although temporary, feature of the transitional landscape of Central Europe, where re-construction of landscape has its most literal meaning.

---

3 Often second-hand cars privately imported from Western Europe, and many of these cars not in the best mechanical condition.

Figure 5.5 Motorization on pedestal: car park in Wrocław, locally known as 'St. Mary of the Cars', 2006

Civic landscapes discourse

The economic transformation of Central European urban landscape has been supplemented by social transition. New landscapes have to respond to many diverse negotiations of reconstructed civic society, their aspirations and complexes. Difficulty in the transformation and reorganization of post-socialist urban landscape expresses the obstacles, problems and aspirations of local societies. New challenges to the rapidly transforming structures and their organization put lots of pressure on the decision makers. The transformation of attitude towards public spaces, landscapes and ownership has been an extreme: from abstract communal/state/Party's/nobody's–everybody's to private and very objective ownership. Restitution of land and building to the former owners accompanied a general defence and unpopularity of public ownership of space. It is the 'shift in emphasis from public to private responsibility: from the nationalization in the collective interest to privatization and commercialization' (Ashworth and Tunbridge 1999, 106). The alliance of commercial powers and urban politics constitutes today's spatial production system of the city. The strategies to revitalize and advertise the city are seemingly carried alongside the need to control diversity by presenting a fully consumable image. 'Private corporations increasingly control the city and its commercial cultural landscape by taking local politics into their own hands. This certain type of manipulation of urban politics is the result of problematic economic situation of many city councils, and lack of sensitivity towards the city as a complex vital social environment' (Cupers and Miessen 2002, 24). Local governments of Central Europe are more vulnerable and lack the experience of the Western cities, so it seems easier to manoeuvre municipalities and enforce the private solutions. According to Kurczewski (2007), since 1989 we have been facing emancipation rather than transformation. Something has been revealed, manifested, liberated, has taken new life and form. One of the first emancipated groups were semi-legal currency exchangers, followed by many others, like capitalist, Catholic, and ethnic and other minorities. There was also political and economic emancipation, as well as cultural and, for many of us most importantly, personal. The possibility of having a passport in our own drawer manifested the most obvious liberation and emancipation, and brought the feeling of freedom.

The assembly of urban landscape managers, defined as persons that influence the visual structure of the city, includes mainly architects and city planners, high city hall officials, and sometimes influential media or NGOs. Dispersion of the decision-making process brought new, powerful landscape lords, including historical preservation offices, city development bureaux and investors or employers, who could, for the sake of a few hundred new working places, often quite freely interfere with the landscape. Landscape had been suddenly liberated from socialist system limitations, and many new landscape decision-makers were ill equipped and unqualified to sustain market pressure and temptations. Local and regional governments, together with planners, were challenged by various pressures and demands. Investors, including local and small and global and powerful, wanted to earn as much as possible on an often speculative real estate market. Local organizations and inhabitants frequently organized themselves to protect the common spaces from total marketization.

Public landscapes were affected by constant struggles and competitions between commercial and civic rights.

Newly found aspirations and expectations facilitated transformation of cultural landscapes. Emancipated societies wanted as soon as possible to be and live 'just like in the West'.[4] This desire for 'Europeanization' constructed new identities. These identities were visualized via sets of symbols, while changing names was the easiest and cheapest way to feel 'almost like in the West'. The new and freely chosen places, shops, streets names can tell a lot about the new social identities; about their aspirations, ambitions, and dreams. Manhattans, Madisons, Wests, Aspens, Beverly Hills or Arcadias facilitate objectives and hopes for better, dreamt world, and actually often based on media-created landscapes, especially Hollywood films. Another name theme reflecting aspiration is derived from European identities. Almost everything can be Euro-technically-whatever:[5] Euro car park, Euro-shop, Euro dry cleaner or office tower. Another set of names is supposed to sound posh for the aspiring middle classes. Shops called Charme, Aristokrat, Olimp or Arkadia (see Chapter 5, Functional transformation of post-socialist landscapes; and Box 5.1 above) describe borders between us, our identities, hopes and ambitions of the less fortunate others.

New landscape managers play important roles in creating and reconstructing recreate symbolic urban features. The process of civic negotiations is frequently long, complex and complicated. Many aspects of public discourse and interpretation have to be considered. The growing role of education, and especially universities, position them as important urban landscape facilitators. Universities, similarly to their Western associates, turn out to be major social, cultural and economic catalysts of many of the post-socialist cities and their landscapes. Discourse on the University Church of St. Paul (*Paulinerkirche*) in Leipzig is a good example of public and civic arbitration (see Figure 5.6).

---

4 The West was, of course, a rather mystic and desired land, mostly based on media, films and indirect opinions, since not many citizens of the Eastern Bloc were allowed to see the 'West'. Additionally, the imagination of how 'the West' looks like was mainly based on the images of the neighbouring countries, like wealthy West Germany, Austria and Sweden, hardly representative for the rest of 'the West'.

5 For example in some countries of the region 'Euro-renovation' means modernization, including PCV windows, fresh paint and so on.

Figure 5.6 Renegotiated landscape: construction of the University Church of St. Paul, Leipzig, August 2007

Box 5.2 University Church of St. Paul, Leipzig

Soon after the Second World War attempts were made to separate the Eastern Bloc universities from their past. Many of them were renamed and theological and well as medical faculties were closed or separated from the university structures. The transformation of landscape of the University of Leipzig represents the fate and changes of many universities of the region. Established in 1409, it was one largest and most recognized in Germany. In 1953 the university was renamed the Karl-Marx Universität and the university life was characterized by political instrumentalization of science and the restriction of academic self-government. In 1968, the medieval Universitätskirche St. Pauli (University Church of St. Paul, also known as *Unikirche* and *Paulinerkirche*) in Leipzig was destroyed because it was a 'wrong symbol' on the grounds of Karl-Marx University, located in central Leipzig's Karl Marx Square. Later on, the main modern university complex was build on Karl-Marx-Platz (before World War II and now again Augustusplatz), including a 28-floor-tall open-book-like building and a modernist block on site of the former St. Paul's church, decorated with 10-metre high Karl Marx metal work. The Paulinen Church of Leipzig University was probably not too important in the mental landscape of pre-war, secular student society, but the demolition of the church in 1968, as an unwanted symbol, made a symbol out of it, and since then the Paulinen Church discourse has became an important part of anti-communist discourse. New governors of the university, mostly from West Germany, wanted to show their power and their ideologies by rebuilding the church. An architectural competition was completed in 2004 and construction expenditure of 140 million euros for a new campus at the old location in the middle of the city confirmed. Only in summer of 2006 was the grand metal feature of Karl Marx taken out of the main university building on Augustusplatz. There will be a new Aula-Kirche-Gebäude 'Paulinum' (assembly hall and church building), an auditorium maximum, faculty buildings and a cafeteria, while the buildings housing lecture rooms and seminars will be converted, extended and redeveloped (Universität Leipzig History 2006).

Another important landscape trail was connected to the developing of new social features of landscape, unwelcome or simply forbidden in the socialist period. Those features include a revival of bourgeois and aristocratic landscapes, anti-Soviet and anti-Russian features, as well as national and minority heritages. Many various forms and meanings have been constructed to represent the above-mentioned expressions of culture. The rapid development of minority institutions and regional heritage attributes symbolized the accumulated and long waited pleas. The restoration of 19th century houses and palaces meet the demand from the newly rich owners who reclaimed their property. The 'landscape in-waiting', often for more than 40 years, erupted frantically as soon as it was possible. Many derelict manor houses and palaces were turned into private residences or hotels. The glorious German, Russian and Austrian imperial landscapes are being rediscovered, appreciated and accommodated into the local heritages in numerous various locations around Poland (Kraków, Gdańsk, Wrocław), Czech Republic (Prague), and elsewhere.

Post-socialist urban landscape policies and practices varied not only between different countries, but also from municipality to municipality. Long awaited freedom brought lots of expectations and hopes for better, more independent and more efficient landscape management that would incorporate needs, desires and even ambitions of the local population. The popular hopes were to be implemented by a new set of decision-makers, often influenced or/and educated through American institutions, like United States Agency for International Development (USAID) and others. American influence on landscape management can be clearly seen in many cities, including Polish ones. American-style belief in personal freedom was quite often adopted and made as a significant landscape management canon. The development of Warsaw can be seen as a perfect example of such a practice, where tall office towers promptly grew up on Wola, the western part of the city centre. The district administration directly applied the spatial liberty policy, and encouraged investors to build up practically whatever they wanted on the plots they acquired. The former Warsaw Ghetto area has been transformed into the landscape of a typical American city in just few years. The over 20-storey-high office blocks neighbour on declining late 19th century housing and post-industrial structures as well as some 1950s and 1960s blocks of flats. The typical American landscape blend, with vertical domination of modern towers surrounded by rather dilapidated setting can be considered as one of the typical post-socialist landscapes, and a result of a specific and not always intentional, landscape management practice (Czepczyński 2005).

Legal gambling and more or less legal sexual services became an important element of cultural landscape in many of the post-socialist cities. Streets dominated by prostitution can be seen in almost every metropolis of the region, but seems to be particularity visualized in many of the Czech–German border towns, and also in Romanian and Bulgarian agglomerations. Budapest is sometimes called 'European Bangkok' (Adamik 2001). Both sex and gambling might be interpreted as a 'shortcut to happiness', the easiest and fastest way to achieve 'the promised arcadia'. Casinos, jackpots, lotteries and other less legal 'easy' financial schemes heat imaginations and help to keep dreams alive. A post-socialist landscape of easy and substitute happiness is additionally enhanced by sex and gambling-oriented tourists from Western Europe.

Democratic and civic practices have their spatial representations. Urban arrangements are re-contextualized or modified in course of practising democracies. There are many modes for how new civic responsibilities are negotiated and visualized in cultural landscape. According to inductive methodologies, some rather minor signs or micro-practices can represent much greater processes. One of the most popular symptoms of social transformation of post-communist landscapes might be development of bike paths. Generally speaking, the development of bike paths is almost always a bottom-to-top procedure: active, local, usually young groups pressure decision-makers to invest in bike path constructions. The significance is greater if the bikes are used as an everyday commuting mode, and not only a recreational activity. In many municipalities, the marked cycling road system can speak of openness of local governances and their will to listen to local demands. Sometimes, the bike paths are made just for the sake of it or for prestige, as in central Bucharest in 2005, where very narrow paths were painted on the pavements, while none of the few cyclists ever used them. In some communities, like in the Northern Polish spa town of Sopot, bike paths are not enough. There is growing conflict between rollerskaters/bladers and bikers on the seaside paths. In 2005 a group was constituted to collect signatures to demand the construction of a special and exclusive roller path in Sopot.[6]

New demand for a 'festivalization of space' facilitated many central public spaces in European cites. In the post-communist part of the continent, old socialist-forced gatherings were replaced by new voluntary public fiestas, including mass New Year celebrations or various open-air and free-of-charge concerts. Large central squares and city parks acquired new, ephemeral functions and meanings. Marathons, bike rides, local and national celebrations, in addition to religious manifestation[7] re-made cultural landscapes of many cities around the region. The growing demand for new public gatherings reflects a changing style of life, towards leisure economy, where sport and entertainment are crucial components of everyday practices. It might be also the answer to growing social atomization and, not only post-socialist, loneliness in the city.

Public spaces often play important roles in civic and political discourse. New, post-modern and generally democratic expressions of civic practices include public strikes, protest marches and political gatherings, often organized in front of the new centres of powers. Parliaments are now being recognized as the main decision-

6 Not successful until autumn of 2007.
7 Mainly in Poland, including mass Corpus Christi processions and public masses.

making institutions, so the protests are usually organized on squares in front of the buildings, like the 'while town' – a set of tents of protesting nurses in Warsaw or the famous protests in Budapest and other Hungarian cities in autumn 2006. Public places and squares, like Budapest Kossuth and Szabadság Squares, were the scenery of mass gatherings in September 2006 to protest against the government. Those places or landscapes are especially important *agoras* in turbulent times of political debate and transformation, when manifestations became important rituals and ways of negotiating. The significance of new public civic landscapes of discourse has also been recognized by politicians, who decided to completely close Szabadság Square in Budapest to prevent meetings, while another hundred 'nomads' lived in a tent city near Kossuth Sq. to prevent the police forces overtaking 'their' spaces of manifestation (see Figure 5.7).

Figure 5.7 Civic landscape discourse in action: 'tent city' on Kossuth Sq., Budapest, 2005

The upheavals in 1989 in various parts of the old communist Europe brought a resurgence, on a new scale and with a new intensity, of nationalism and territorial parochialism, characterized by claims to exclusivity, by assertions of the home-grown authenticity of local specificity and by a hostility to at least some designated others (Massey 2006). In 2006 in prime ministers of the Vishehrad Group met to celebrate 15 years of the organization. During that meeting, the Czech Prime Minister said that 'after 15 years one has to admit, that the way to the dreamt Western style of life is not as simple, as we thought'. The disappointment seems to be quite overwhelming and encourages populists and extremists all over post-socialist Europe. Political parties like the Polish 'Samoobrona' (Self-defence) or Liga Polskich Rodzin (League of Polish Families), Hungarian Party of Justice and Life (MIEP), Bulgarian Ataka (Attack) or Partidul România Mare (Party of Great Romania) actively promote xenophobic, homophobic, conservative, revisionist programs to increase popularity. Anti-Gipsy (in Czech Republic, Slovakia, Hungary, Bulgaria and Romania), ant-gay (mainly in Poland), anti-Turkish (in Bulgaria), anti-German (in Poland and Czech Republic), anti-Jewish (for example in Poland and Hungary), anti-Hungarian (in Slovakia and Romania) and general anti-stranger attitudes marginalize and exclude a part of society from the national discourse (Guillemoles 2007). These positions are also visualized in urban landscape; partly as manifestations of local policies (see more Chapter 5, Landscapes of the excluded), but also as the significance of the minorities' institutions and spaces. Conservatives mayors opposed 'Gay Pride' and 'Love' Parades in Warsaw and Poznań, and often promote nationalistic and controversial icons and symbols, like the mythic Hungarian bird *turul*, symbolizing historical pride, but also to some extent militarism and revisionism. The rise of xenophobic and nationalistic narrations is visualized in graffiti, used as a way of public communication, mainly aimed against alternative social groups. Sometimes chauvinistic local leaders can limit local development, preventing certain nationals from investing. In the early 2000s, an ultra-conservative, populist and 'Germanophobic' mayor prevented or voided German investments in Szczecin, north-western Poland.

The revival of spiritual landscapes is not the most characteristic of every country of the region. The power of religious landscapes is most obvious in Catholic Poland, and to lesser extent in Bulgaria and Romania, while in industrialized and secularized Czech and East German societies religion seems to play a marginal role in urban cultural landscape. The church and its institutions became a part of the official landscape, supported by national authorities. Long awaited freedom of religion brought the return of official ceremonies, incorporated into national schemes. Huge, modern churches can be seen in almost every neighbourhood block of flats in Poland. New churches represent a new revival, often not of real faith, but of empowering institutions. Long and still uncertain construction of the Divine Providence Temple in Warsaw, mainly financed by public money, reflects the ambiguities which accompany spiritual landscape reinforcements. Concurrently, the form and size of Poland's largest church in located in small village of Licheń, central Poland, can represent a conservative and somehow 'Byzantine', but dominant style of Polish Catholicism (see Figure 5.8).

5.8 Power of faith: St. Mary Basilica in Licheń, 2006

Box 5.3 St. Mary of Licheń Basilica

The largest church in Poland and the eighth largest in the world was completed in 2004, ten years after beginning construction. This five-nave Byzantine-style basilica was constructed under supervision of influential priest-manager of the local late 19th century St. Mary of Licheń sanctuary and pilgrimage centre. The complex was designed by Polish architect Barbara Bielecka. The church is 139 metres long, with 23,000 square meters of floor and a 141-metre tower, while the main cupola is 25 metres in diameter and 45 metres high. Thirty-three steps lead to the cross-planned 'symbolic building for 21st century', decorated with 365 widows and 52 doors. The main portico is dominated by the massive sculpture of the Queen of the Angels with the Infant, accompanied by six giant angels. The church, according to the official website (Bazylika Najświętszej Maryi Panny Licheńskiej 2007) is supposed to be reminiscent of fields of gain, and the grain motif is repeated in gold-amber window glass and other 'grainy' ornaments. The central part of the basilica floor is decorated with 40,000 pieces of marble mosaic, while the lower church is dedicated to 108 Martyrs of the Second World War. This most controversial place of worship in Poland is for some symbol of kitsch and trumpery, while for others 'connects all the best and classical European church architecture' (Bazylika Najświętszej Maryi Panny Licheńskiej 2007). The church construction and operation has been financed by private and institutional donations, but also some state-owned companies.

Since the early 1990s Poland has faced growing cult of the 'Polish Pope' – John Paul the Second. During the life of the Polish Pope dozens of street names and square were renamed, and hundreds of monuments and plaques were dedicated to the Pope. The process of cultural landscape conversion caused by the cult of the Pope has accelerated after his death in 2005. The course of 'John-Paul-the-Second-ization' of Polish cultural landscapes includes naming a John Paul the Second Catholic University in Lublin, a John Paul the Second Pier in Sopot, together with many schools, hospitals and other institutions. Every anniversary of his death is celebrated on main squares of almost every Polish city: thousands of candles are left to commemorate the late Pope (see Figure 5.9).

Figure 5.9 Religion and cult: murals on the Zaspa estate, Gdańsk, 2006

One of the main differences between urban societies of 'old' and 'new' Europe concern the accepted spectrum of behaviours and lifestyles, respected by general public. By and large, the 'be rich or die' rule dominates in the post-socialist countries (Nawratek 2005). Often people who do want mainly to be rich, or are less fortunate, are excluded and left to themselves. The public spaces slowly become mainly for the 'young, rich and beautiful'; the squares and parks are sold to private developers; the number and size of gated communities can only be compared with some parts of southern California. The state and the local governments are losing interest in managing real public spaces, often conveying the responsibilities to shopping mall developers. The malls, with post-modern *agoras*, fountains, pseudo-streets and highly secured and monitored environments, although formally open to everybody, are often not really accessible, open and friendly for the poor and less fortunate. In a practical sense, public spaces, more and more limited by legal decisions and architectural barriers, are an alternative for all assortments of quasi-public spaces of hypermarkets, shopping malls and theme parks.

Urbanization processes

Urban landscape forms seem to follow three main directions: either they continue a traditional appearance, which usually brings conventional and predictable connotations; create new, avant-garde forms to surprise recipients, and impose, by surprise, new meanings and subtexts; or as often seen in Germany, they are as practical, square and functional as possible[8] (Nawratek 2005). Post-socialist cities, especially large agglomerations, are challenged by contemporary trends in urbanism. For many local and regional municipalities and societies new, post-socialist cities must be shining and competitive, also in visual appearance. The glamourization of space and landscape is probably most clearly visible in shopping malls, but also in new apartment buildings. Central European cities seem to be more vulnerable and unprepared to confront rapid transformation. Many of the post-socialist cities, especially in Poland, Bulgaria and Romania, seem to continue a late 19th century, American-style urbanization pattern. Probably the most similar is the relationship between investor and local government, where a strong and technically omnipotent investor dictates the rules of urban landscape. Municipalities are practically ready to fulfil almost every with of the investor in order to bring the investor to the city.

Consumption, including 'visual consumption' of landscape and market demand for cultural landscape features, has created a frame for further transformations towards, as Ritzer (2007) puts it, 'globalization of nothing'. 'Best' practices and designs copied from some successful cities reproduce icons of blue glass and steel, reproducing 'no-places' that could be anywhere. Freedom of competition brings vigour, but also confusion and disorder. New functions and features emerge in fast developing cities. Financial institutions, advertisements, office towers, and media have reshaped and changed the form of the cities, idiosyncratically chasing 'the

---

8 Nawratek (2005, 201) teasingly relates this type of architecture in an advertisement for German chocolate: *Quadratisch – Praktisch – Gut* (Square – Practical – Good).

better 'West'. The development of Warsaw can be seen as a perfect example of such a practice, where tall office towers promptly grew at the western part of the city centre. The district administration directly applied the spatial liberty policy, and encouraged investors to build practically whatever they wanted on the plots they acquired. The area has been transformed into the landscape of a typical American city just in few years.

There is a growing market pressure in many of the post-socialist cities. Recently Prague has faced new high-rise development plans. The southern part of the Czech capital, Pankrác, is supposed to become modern Prague City. Three already-existing high rise towers are going to be accompanied by a V-shaped 31-floor apartment tower, and a 21-floor cyllindrical hotel. The buildings have been accepted by the local historical monuments warden and National Heritage Institute. Many local inhabitants are opposed to the idea, since it destroys the historical landscape of the city. UNESCO experts believe that the World Heritage listed city and its vistas will be spoiled by new high-rise developments. Prague clearly wants to complete with Warsaw, Bratislava, Berlin and Sofia, developing new office space and meeting growing business demand (see Sýkora 2007). There is a growing demand and supply of the apartment towers, perfectly located within older housing stock close to the city centres or with perfect view. Many of the towers have English, or rather American names, like 28- and 36-floor-high Sea Towers in Gdynia, and the 56-floor Sky Tower in Wrocław. Some of them, including the newest plan for the 51-floor Big Boy tower in Gdańsk, are located on the edge of a huge 1970s blocks of flats housing estate (Murawski 2007). On the other hand, rather unique, post-/ anti-modern often megalomaniac and massive forms might indicate aspirations and global tendency and search for iconic buildings.

In some post-socialist countries, especially in Poland,[9] urban landscape continuity has been torn by the development of gated communities. This form of residential estate is characterized by a closed perimeter of walls and fences, containing controlled entrances. Gated communities usually consist of small residential streets and include various amenities. For smaller communities this may be only a park or other common area. For larger communities, it may be possible for residents to stay within the community for most day-to-day activities. Gated communities are a type of common interest development, but are distinct from intentional communities (Lewicka and Zaborska 2007). Though they are called 'communities', there is no evidence to suggest that social capital is any higher within them than other forms of residential development. Given that they are spatially a type of enclave, they are more likely to have negative contributions to the overall social capital of the broader community (Low 2001). Some gated communities, usually called 'guard-gated communities', are staffed by private security guards. These communities are often home to high-value properties, or set up as retirement villages. Some gated communities are secure enough to resemble fortresses. Gated buildings disintegrate urban space and gated communities seldom turn out to be real communities, but rather a selection of socially isolated individuals (see Lewicka 2005, Gądecki 2005). 'In our gated communities and shopping malls we might enjoy the clean imitation of

---

9 And also in a much more developed form in Russia, especially Moscow.

reality, and of course the reality will always be inferior to its cloned mirror image' (Cupers and Miessen 2002, 26). Different social constructions are related to the gated communities: for some they are safe asylums in modern, vicious cities, while for some others gated estates are merely 'ghettos of the rich'.

Box 5.4 Marina Mokotów, Warsaw

A prestigious residential estate is being constructed in the green area at the corner of Żwirki i Wigury and Racławicka streets in the southern part of Warsaw. It is developed by a company set up by PKO Inwestycje – a firm owned by Poland's largest bank, and by one of the largest developers, Dom Development SA. The construction started in 2003 and was partially complete by 2007. According to the investment project, Marina Mokotów estate is to be self-sufficient. It is to exist within a city and take advantage of its assets including a location relatively close to the city centre, next to one of the city's most important streets – Żwirki i Wigury, which runs to the Okęcie airport. At the same time, it is to be 'an oasis of peace and quiet in a busy metropolis that will provide its residents with comfortable living and relaxation-recreation areas, shops, restaurants, a post office and a bank'. The ground floors of buildings earmarked for these functions will feature 5,000 square metres of space. Marina will include a total of 1,500 apartments for up to 6,000 people. The 22 luxury apartment buildings will hold units of between 37–220 square metres, while the six residential buildings will contain apartments of between 33–131 square metres. There will be nine urban villas of 300–400 square metres plus up to 1,500-square-metre plots. The semi-detached houses will offer from 200–350 square metres and will come with plots of up to 700 square metres (Marina Mokotów 2007). Marina Mokotów is the first large estate of luxury houses implemented in the immediate vicinity of the city centre. Since the very beginning, Marina has been planned as a gated estate, a town within the city. This idea is reflected in its layout. The main axis, along which the buildings are arranged, is formed by a man-made lake. Apartment buildings are clustered around this central part. The estate has new streets, public squares and a lot of greenery, including a park arranged on 19th century fort embankments (Piwowarczyk 2005).

Shrinkage of the cities is a complex phenomenon, apparent in every post-socialist country of Central Europe. It is not just being driven by falling population numbers and growing amounts of empty housing stock. Extensive industrial wastelands, abandoned former military sites, empty schools, kindergartens and other community facilities, markedly underused technical infrastructure, a drastic collapse in public revenues and deep cuts to public social and cultural provision are also having an impact. Shrinkage means the functional under-use of the whole urban structure. It drains away the vitality from urban areas. This process has mainly been caused by the radical economic and social changes that have occurred in the course of the transition from one social system to another and in German reunification. In addition to this, some of the political decisions taken in the early 1990s helped to exacerbate structural problems and difficulties in the housing market. Renovated but empty houses in many East German towns speak about much more difficult transformation:

depopulation caused by mass migration and accelerated by negative natural growth (see Oswalt 2006, Taverne 2004).

The challenge is to initiate a comprehensive structural adjustment that opens up the city again in future as a place for people to live, spend their leisure time and work, and to do so under conditions characterized by limited private economic power and extremely tight public finances. With the Urban Restructuring in the New Federal States programme, for the first time in urban development, legal, planning and economic instruments have been applied to systematically exploit a shrinkage process, with shrinkage being understood as a combination of enhancement and demolition (see Wallraf 2006). The close linkage of structural adaptations pursued by means of urban design and adjustments to the housing market is also new, although the concerns of sustainable urban development are supposed to have priority over the interests of the housing business. In practice, East Germany's cities are pursuing various restructuring strategies (Oswalt 2006). They extend from conceptually more or less well founded clearances at individual points within the overall structure of the city to measures to thin out intensively built up areas using the tools of urban planning and the mass demolition of whole quarters. Grass is the only thing growing on some demolition sites, but others are being redeveloped with new forms of residential accommodation or parks. Apart from the low-cost variant of total demolition, the options include expensive redevelopment activities centred on 'beacon projects' that create a completely new image for a location (Wallraf 2006). The problem of shrinking cities, although most significant in East Germany, appears in every country of the region, especially in the old industrial zones, like Bytom, Wałbrzych or Sosnowiec in Poland, Ostrava in Czech Republic, but also in some smaller, and deindustrialized towns like Vsetin in Czech Republic or Plovdiv in Bulgaria.

Revitalization, reurbanization, regeneration, and gentrification of inner city landscapes are tendencies reinforcing each other. Hundreds of historical city centres have been renovated and modernized to respond to a growing social demand for historical landscapes. Revitalization of old, inner-city houses and infrastructures (see Figure 5.10) have resulted in, on one hand, better utilization and higher revenues, but on the other hand, in the growing erosion of public life by new capitalist power. Regeneration programs were aimed at gentrifying the city centre, increasing quality of life, and enhancing tourism, which was often the main goal and priority. Some, considered to be the most successful revitalization projects, like Old City in Kraków or Stodolny Street in Ostrava, have become practically mono-functional, mainly bars, pubs and entertainment districts, somehow reminding one of a *simulacrum* of a real town. Probably the largest redevelopment and revitalization project is Budapest Millenáris Park. The park itself is actually the reconstructed Ganz Electrical Factory and one can still see the parts of machinery that once stood here. Today it is it just a park, but also a scene for exhibitions, plays, concerts and performances and a modern recreation centre with exhibition halls, underground parking lot, playground, park, ponds, café and theatre.

Figure 5.10 Revitalization at work: central Košice, 2007

The *archaization* of landscape is a tool or method of assimilation of post-communist spaces, relating them to the 'happier', pre-communist times. Historical references in urban design, architecture and cultural landscape are always political projects, where somebody – an investor, developer, municipal official or buyer, chooses a preferred historical period to be copied or faked. This process is connected with the demand for historical legitimacy. *Archaization* and the popularity of neo-pseudo-historical landscapes answers the social call for anti-modernist, as well as verified and proven, solutions. History has become one of the strongest points of reference and the most reliable anchor in post-modern, turbulent times. Learning and discovering one's own, often forbidden past is connected with learning from best solutions and can been seen as re-constructing, copying, and faking chosen historical forms. Many contemporary buildings, both public and private houses, resemble to some extent historical buildings, and many of them were built as more or less accurate copies of old structures. After 45 years of enforced communist 'dictatorship of landscape', the post-socialist societies turned to the local, regional and national icons, myths, histories and symbols. A special role was played by new right wing, nationalistic parties and governments, which anchored their identities in glorious, pre-communist pasts and allegories. The affirmation of historical glory was followed by reconstructing of national icons, like Jabłonowski Palace in Warsaw (Figure 5.11), the reconstruction plan for the Berlin City Castle (see Box 4.3) or Warsaw's Saxon Palace. Ironically, many of the urban development of the 1990s are reminiscent of the 1950s schemes implemented during the communist period in Poland (see Chapter

3, Historical Policy of Landscape). The inner city is often preserved, while most new investments are suburban greenfield developments (Nawratek 2005).

Figure 5.11 Archaization of landscape: new Jabłonowski Palace, Warsaw, 2005

EU structural funds have also directly and indirectly influenced the transformation of urban landscapes of Central Europe. Many projects are being financed to create new workplaces, to limit negative aspects of unemployment and to enhance local economic development. Some of those projects are based on historical aspirations and potential tourism development. Historical references seem to be most popular in search for originality and uniqueness, sometimes to rather extreme scale, like the 'reconstruction' of an early medieval fortified village – theme park in Byczyna, southern Poland. This Polish-Czech Knighthood Training Centre was completed in September 2007 and comprises a smithy, armoury, inn and pottery workshop designed to 'integrate and liquidate mental barriers between Czech and Polish population' (Piński and Piński 2007, 67). Another rather peculiar project, initiated to obtain EU money, is located in Pruszcz Gdański near Gdańsk, and is based on the 'revitalization' of ancient Roman camp, located on an old amber trade trail. A 'copy' of the original Roman camp, called *Faktoria*, is planned to be developed. The project, partly financed by the EU regional development funds, aims to 'reconstruct' the ancient settlement as it apparently was in the 2nd century AD. This 'living museum' will cover about 1.5 hectares on the banks of the Radunia river and three man-made islands, and will house craft workshops, including goldsmiths and amber workers, in addition to houses and farms. The camp will accommodate amber festivals, fairs, concerts, museum and many other cultural and sporting events. *Faktoria* will mark the beginning of the *Via Ambra*, or Amber path, connecting the Baltic coast with Italy.

De-urbanization and suburbanization are new aspects of post-socialist cultural landscapes. Rapid urban sprawl changes not only suburban villages and communities, but also urban centres and has created high transportation demand and traffic jams. Fast growing immense suburban zones of Warsaw, Budapest and Prague can probably only be compared to American suburbanization (see Gądecki 2005). The vast majority of new suburban developments continue well-established aesthetic trends. There is a domination of single family traditional housing, built of brick and red-roofed. Derelict shops in the city centre of Budapest represent a functional transformation and shift of the shopping culture. Investments located on the city ring are economically more feasible and accessible for a motorized, suburban middle class. Many but hardly local development strategies and policies are focused on encouraging the wealthy to move into the city to pay higher taxes, and the same time force the poor and their problems outside the city (Nawratek 2005).

Landscapes of the excluded

Soon after the transformation began many negative aspects of openness to global markets and trends could be noticed in many post-socialist cities. Large social groups have hardly benefited from the economic freedom: the old, the ill, the less entrepreneurial and the passive have been almost forgotten by the politicians. The rapid rise of organized and not-so-well organized crime transformed the social sense of security. Many global players came with global capital, but have not always played fair with the (to some extent) naïve and unprepared local and regional companies. Social and moral liberations, following the 1968 Western contestations, did not occur in the Socialist Bloc until the early 1990s, when they shocked traditional families and values. 'Today's upscale, pseudo-public spaces – sumptuary malls, office centres, culture acropolises, and so on – are full of invisible signs warning off the underclass 'Other'. Although architectural critics are usually oblivious to how the built environment contributes to segregation, pariah groups – whether poor Latino families, young black men, or elderly homeless while females – read the meaning immediately' (Davis 1990). Today's urban environments produce an easy, self-conscious identity by throwing up an image of who belongs in the city and who does not. By attracting only regular users, space for different minority groups is increasingly threatened. The concentrations of exclusive malls, upper-class gated districts and CCTV-controlled streets and squares turn the other streets and districts into a 'no-go area'. Social polarization is spatially reflected, and this reflection is clearly visible in urban landscape and its forms, functions and meanings (see Domański 2002). In many post-socialist cities the centres have been inhabited by poorer social groups, sometimes even by the socially marginalized. The situation is a direct consequence of the communist forced lodging policy, where, instead in social housing, people were located in old, bourgeois quarters, where original often high-quality flats were divided into many smaller ones, with shared corridor or bathroom. Recently, according to new gentrification plans, the poor from the city centres are being pushed by developers and local governments into isolated peripheral districts.[10]

The polarization of the Central European urban landscape is strengthened when the economic divergence is added to the ethnic one. Most of the countries of the region[11] have a burning and growing problem of Gipsy/Roma minorities. The conflict is probably most obvious in Slovakia and Czech Republic, but also in many Bulgarian and Romanian towns and cities, where the most drastic steps were taken to separate the excluded Gipsies form the 'normal' society (see Figure 5.12). They live in grinding poverty, trapped in a vicious circle of poor education, teenage pregnancy, unemployment, petty crime and alcoholism. In scenes reminiscent of the segregation-era South of the United States, the Roma, who tend to be darker-skinned than the majority, are often denied work, housing or social benefits simply because of their skin colour. Roma children are routinely shunted to schools for the learning disabled, and Roma are regularly harassed by the police and often physically attacked by groups of skinheads (see Repa 2000).

Figure 5.12 Landscape of the excluded: Fakultet, Gipsy ghetto, Sofia, 2005

---

10 In some cases, like in the quite wealthy and posh appearing town of Sopot in northern Poland, the local government bought a barrack in a suburban commune, outside the town limits, to resettle those who cannot/do not want to pay rent.

11 Only in East Germany and Poland have the Roma or Gipsy minority not became a major social issue. All of the other countries of the region, including Czech Republic, Slovakia, Hungary, Bulgaria, Romania, but also countries of the former Yugoslavia, have numerous and generally excluded Roma minorities.

The most spectacular examples of landscape exclusions were accelerated in 1998, when local governments of two Czech industrial cities, Usti nad Labem and Pilzen, decided to separate 'the good' and 'the bad' with walls. The two Czech cities have decided to fence in what they call 'problematic' public housing residents, creating what is virtually a ghetto for the residents, mainly Gipsies, who officials say ruin the calm, orderly life of their neighbours. City officials say fencing in such citizens, and guarding them with round-the-clock police patrols, is the only sensible way to deal with people who refuse to pay rent on their city-owned apartments, throw garbage into the street and gather on sidewalks talking, singing and sometimes drinking until late in the night (see Green 1998). Czech opinion polls consistently suggest a large degree of popular hostility towards Gipsies. Some local authorities have even tried restricting their movements, only to be overruled by the central government. But after decades of centralized communist dictatorship, the courts appear keen to enforce the principle of limited government and a system of constitutional checks and balances (Repa 2000). Many Central European municipalities face similar problems of cultural, social, economic and spatial exclusion of the Roma minorities. Two most popularized cases of segregation happened in block of flats or *panelák* housing estates in Chánov in the towns of Most and Maticni in Usti nad Labem in the north-western part of the Czech Republic.

Box 5.5 The Wall, Usti nad Labem

The wall was built in 1998 after complaints by locals that the Gipsies were noisy and unhygienic, but criticized by human rights groups and the European Union as racial segregation. The Gipsies had mixed feelings: some were celebrating, drinking rum, but they also said that the town council's plans to buy the houses of other local residents, enabling them to move away, basically amounted to the creation of a Gipsy ghetto in the street (Repa 2000). 'The fence will separate this problematic community from those people who have private houses on the road', said Milan Knotek, spokesman for the Usti city hall. 'The wall will not stop them from moving about. It will not be a ghetto enclosed on four sides'. 'This is a concentration camp', said Pavel Dostal, a Social Democratic member of Parliament. In Usti, the non-Gipsy residents of Maticni Street say they are not racist, they just want to be left alone. Hana Chladkova, a 27-year-old insurance clerk, says the wall was her idea. She says she just wants to sleep at night and let her three-year-old son, Petr, play in peace. Instead, she says, the noisy neighbours, the garbage strewn around their apartment block and the rats that come with it are too much. 'A barrier will be more aesthetic and it will keep out the noise, the dirt and the stink', Mrs Chladkova said. Several said they were illegally resettled there, and they say the city fails to offer the services, from hot water to garbage pickup, that they pay for. Jan Kocourek, the deputy mayor of the Usti district that includes Maticni Street, defended the plan to build a wall. 'These people have a different way of living, from afternoon until late at night, and they create noise', Mr. Kocourek said (Green 1998). The Czech government, which opposed the wall from the start, provided the local authorities with state money for social welfare programmes in the town, but much of the money will be used for buying up residents' houses and for removing the wall. In October 1999, the Czech parliament ordered the wall to be taken down – and the council complied. Some non-Gipsy residents are reported to have been offered alternative properties. In its ruling, the court upheld

Parliament's right to issue a moral condemnation. Although the wall has gone, the ruling could inhibit the central authorities in Prague from intervening in controversial local government decisions in future (Repa 2000).

The poor, unemployed, unsuccessful or 'project-less' have been most often left to themselves. During the communist era the egalitarian policy was aimed to provide basic means of life for everybody, irrespective of talents and efforts. Now some of the governments try to ignore the poor, while some others try to cope with the brand new situation. Peripheral landscapes are usually located in low-income suburbs, but sometime also in central, but substandard inner city districts, like in Łódź in central Poland, where the district of former Jewish ghetto is a typical leftover space, a 'non-site'. These deprived and semi-derelict areas are usually clearly detectable by smell, dirt, lack of infrastructure and renovation.

Chapter 6

Interpreting Landscapes in Transition

The turn of the 1980s and 1990s was marked by great changes in Central Europe, simultaneously stimulating expectations of a new and better future for the newly liberated nations, which sooner or later initiated processes of democratic transformations. From the perspective of 18 years, it has become evident that these processes will continue for many years to come. There is no overriding direction to global culture; the market prevails and demands continuous differentiation. Most of the dominant 20th century metanarratives are in decline: communism has lost its adherents, religions are losing authority,[1] the belief in economic and technological progress and democracy do not supply satisfactory answers for the main social problems of the early 21st century. A general absence of a deeply held metanarratives is one of the major problems (Jencks 2005, 203). One of the young persons interviewed by Woźniczko (2007, 44–48) said that 'we lack any idealism now, despite all the negative aspects of the Soviet Bloc, one has to say one thing: it was the land of the idea. Now all we have is the objective of consumption'. Another, 40 years old, noticed that 'in the early 1990s the country was poured by an avalanche of consumption goods: jeans, sausage, sugar appeared to be much more important than any idea. Now we look for new ideas that can become ideologies'. Christianity, Marxism, glamour, and nationalism are possible ideas to follow. Many are still looking for someone who will show them the 'right' and only way. After some years of de-ideologization of cultural landscapes, recent years have brought, especially in right-populist governed Poland, a new wave of re-iconography. Conservative and national-Catholic icons appear in many Polish cities. The return of a new ideological landscape is visible in many cities around the region, often as a counter reaction against globalization. More recently, national politics are becoming more and more ideological. There are many attempts to overtake public space by the ideological right or conservatives, especially in Poland and Hungary, followed by regulations about whom and when can use the space (see Chapter 5, Civic landscapes discourse).

Contrary to popular theories and imagination, people are becoming more and more uniform and alike (Figure 6.1). Wherever we go, we see the same modern products, cars and buildings; people are dressing the same and have more and more similar behaviour. The remains of the old times: traditions, historical monuments, together with the natural environment, become the only differentiating factors. Despite the notion of freedom and countless alternatives, our choices are similar or the same; we buy the same products, our actions and manners are more and more standardized (Wagenaar 2004). Many sociologists and philosophers, like Ortega y Gasset (1994) say that modernity is not a time of individuals, but of masses. Individuals are more

---

1 Except for Islam.

and more controlled and dependent on each other and on the system; as a result they become one mass. Ortega (1994) suggests that mass society is based on the belief that being too distinguishing is indecent. Jean Gottmann (1990) identified two major bundles of opposing forces, which animate geographical space: circulation and iconography (iconology).

- *Circulation* de-partitions space, opening it for the best and the worst by churning together individuals, goods and ideas. It engenders universalism and cosmopolitanism. Circulation brings about changes, generated by global flows, movements and so-called modernity. The concept of circulation is based on the Aristotelian idea of expanding Greece and the pluralist system and the power of multilevel communication. Dynamic development, economic priorities and expansion rather than equilibrium attracted many societies and presently seems to dominate the globalizing landscapes of Central and Eastern Europe.
- *Iconography* resists movements and partitions the space. This force, more abstract then material, rests on identity and symbolic links. Through their iconography, groups share the same representations, visions of the world and values, uniting them within common space of belief. Iconography creates stable identities and helps to maintain theses identities by resisting generalized circulation and by partitioning the space. Icons carry a meaning, which they bestow on those places where they provide roots to people. Icons offer an image of the world as much as they make of the individual self in the world: they are a worldview from a particular standpoint. Inspired by Platonic thoughts, iconographical communities try to protect themselves from foreign influence or any force that could engulf them and keep themselves well balanced, protected and safe (Bonnemaison 2005, 43).

The circulation/iconography dialectic facilitates the understanding of many of the tensions and controversies currently occurring worldwide: opening versus closing, universalism versus localism, cosmopolitanism versus isolationism (Bonnemaison 2005). European nations are balancing between the need to build a greater entity and the resistance of established national and regional iconographies. New cultural geography looks at the construction of collective identities and their spatial 'territories', while cultural landscape studies decode, interpret and elucidate the relations of forms, functions and significance. The struggle between flows and icons are extremely visible within transitional society. The change imposes and generates choices to be made, and prioritizes one paradigm over another. Post-socialist societies and countries are probably among the best modern examples of that tussle. The circulation and iconography discourse has been a part of the ideological conflict. The cultural landscape has been employed and manipulated as tool and weapon by most dictatorships, especially by 'the dictatorships of proletariat' (Czepczyński 2006a).

Figure 6.1 Emancipated landscape: Warsaw city centre, 2007

Today we understand that universe is not, as Einstein thought, an eternal cosmos, but rather a process of cosmogenesis, that is an expansion and complexification, one that roots us in time and cultural space, one that orients a global culture grown weary of past dogmas. This narrative shows creative destruction and nonsense; it shows real chaos and wandering. But also it reveals 'increasing organization, sensitivity, and further complexity growing out of chaos. It reveals the way we are typical cosmogenic beings' (Jencks 2005, 210). Production, constant readjustment, and restructuring of cultural landscape in the contemporary city become the permanent, perpetual mechanism and integral part of our everyday life. At the same time the post-socialist landscape is compared with the previous, socialist one. Constant contextualization and assessment with the past, the West and possible future facilitate much of the modern landscape discourse. Post-socialist cities are post-socialist not because they are better or worse then any other cities; they are post-socialist in the sense that they are different from other cites (Harloe 1996). Tracing the differences and dissimilarities between non-socialist[2] and post-socialist urban landscapes and development patterns have been core foci of this synthesis. The dramatic character

---

2 It should be remembered that many of the West European cities were, to some extent, quite socialist, and governed for decades by left, socialist leaders, and now those cities feature many socialist landscape marks. One of the best examples is Vienna, with hundreds of clearly marked social houses and other signs of social/workers' power, while black eagle in the Austrian coat of arms still carries the sickle and hammer in its claws.

of public space implies the presence of a theatrical stage. The built space of the public sphere or urban landscape can be seen as a theatre set, and specific social and architectural ideas can be applied to it (Cupers and Miessen 2002, 36).

Settings of Central European cities carry stigmas of half a century of state socialism. One of the worst post-socialist heritages is the 'destruction of natural spatial and social relations, and cities are not seen as societies in spaces, but rather only as a congregation of buildings' (Nawratek 2005, 188). The reconstruction of the civic significance of urban space is among the most important tasks in front of post-socialist societies. The production of new layers of meaning and different interpretations of post-socialist landscape is an ongoing process. After the rapid conversions of the early 1990s, the process of landscape reinterpretation has recently either settled down or entered another phase. De-communization and transformation of meanings are always connected with cultural background of society, as well with aspirations and hopes. Landscapes of Central Europe are liminal in an essential sense of the word (Turner 1975): not socialist any more, but still not truly liberated from the old traumatic and totalitarian burdens, represented to a great extent through the situation of regional societies, sandwiched in between things they want to remember and things they would be happy to forget. Each above-mentioned icon is loaded with layers of meanings, texts and connotations, attached to it by various social groups of decoders. All of the existing processes of social and urban conversions are facilitated by local and national powers and memories. The old and new semantic rulers transform the old icons via media, law and money, but despite their intentions and ambitions, the meaning of cultural landscape is always verified by everyday users, who provide the real significance. Our relation to the socialist past is a mixture between this we remember, remind and want to recall and forget.

The destiny and prospect of old socialist iconographical landscape features, like monuments and communist party headquarters, can be seen as a peculiar 'litmus paper' indicating the fears, ambitions and aspirations of post-socialist societies. De-communization and transformations of meanings are always connected with the cultural background of society, its history, structure, wealth as well as aspirations and hopes. Attitude towards post-socialist icons mirrors preceding humiliations and dictatorships, as well as present acceptance and reconciliation with their own history and can be seen as an explicit indicator of political and cultural transformations. The same time fate of old communist symbols represents attitudes towards the 'recent past' and can indicate the position in the process of liminal transformation. The past manifested in memory practices of commemoration and rejection influences contemporary identities and, to a further extent, future opportunities and developments. Cultural landscapes, as mélange of forms, meanings and functions, project and represent the character, spirit and eccentricities of society. Some social and cultural groups create their own systems of representations, based on a distinctive construction which results form particular experiences and expectations. The attitudes towards post-communist cultural landscape residua can be grouped in three main schemes or social constructions (see Czepczyński 2006c):

- *Funky*, usually constructed by the young and trendy, looks for new inspirations and stimulations. Combination of leftist icons and the 1970s design result in quite attractive product, appealing to many who never experienced communism themselves. Thematic pubs and bars, full of communist propaganda and icons, as well as gadgets and T-shirts use, re-use and recycle the old symbols in a brand new cultural context (Brussig 2002). The red star, Honecker's picture or Lenin's head are hardly anything more then an aesthetic sign, trendy and fashionable in some of the social groups. Demand for socialist kitsch, de-sacralized and recycled icons seems to be merely an original and visual trend, sometimes and for some people promoting forgotten or unrealized ideas. This construction is mainly focused on the interpretation of forms and style of iconic landscapes, generally unrelated to their former functions and significance (see Chapter 4, Reincorporation and new construction of old landscapes; and *Ostalgic* landscapes and representations).

- The *Freaky* attitude is focused on destructive and critical aspects of the system. The communist period and associated cultural landscape is critically contextualized as a time and space of oppression, devastation and tyranny. Disgraceful and/or insignificant features can only bring the dark memory back, so the 'recent past' and its residua should be removed or eliminated. Remembering mainly negative substance of the state communism by and large prevents substantial reinterpretation of communist landscapes and values. Coded meanings seem to dominate this explanation, visualized in anti-communist memorials and museums (see Chapter 4, Memorializing anti-communism).

- The *Fantastic* construction relies on using and incorporating into functional urban tissue the best of what is left. Due to limited connection with the outside world, the socialist landscape had resisted, to some extend, the globalization flows until early 1990s. There is a growing demand for grandeur and symbolism in the post-modern world, which can be found in many features of the socialist cultural landscape. A growing tourist demand and often limited local attractions force local societies to re-interpret old icons to meet requirements and pressures of competitive markets. Functions predominate this understanding and assimilation of cultural landscape features, while forms and meanings are only supplementary. A new social context of nationalist pride has been attached to the Civic Centre of Bucharest, especially to the Palace of the Parliament, former Ceausescu's People's Palace (see Chapter 4, Reincorporation and new construction of old landscapes).

There is a growing regional schizophrenia and schematic chaos of interpretations, but this chaos might be very fruitful and is in clear opposition to the previous 'only one way of interpreting' experience during the communist era. On one hand, communist designs, films, objects become funny and funky components of everyday life, but on the other hand, official historiography and historical policy present the 45 years of communism as the darkest period of regional history. Many people are lost between official anti-communist propaganda and popular memories and connotations. The communist history presented in mass media consists mostly of comedy films and humorous gadgets. Bookstores are full of critical anti-communist monographs and crime documentaries, together with collections of communist jokes, reprints of old posters and amusing cuts from communist propaganda films. Even the documentation centres and anti-communist museums often sell Trabants and funky gadgets, somehow promoting the 'recent past' (see Pietrasik 2007).

The urban landscape can be of important value to the local society and economy. Cultural landscape can be seen as a feature and function of a place, as well as a development option. Well-managed and consumed landscapes can enhance the quality of life, increase residential, investment and tourism attractiveness. The urban setting equally frames and creates the local milieu. Oscar Wilde, is quoted in an office tower advertisement: 'it is only shallow people who do not judge by appearance' (Dovey 2001). The observer always decodes, interprets and judges urban landscape's meanings and structures, as much as he or she wants and can (Czepczyński 2005a). Foucault suggests that modern power is a dispersed set of micro-practices, many of which operate through the normalizing gaze of surveillance regimes (Dovey 2001). Hundreds of practices implemented by hundreds of small- to large-scale landscape managers fill the modern urban landscape jigsaw. The old and new semantic rulers transform the old icons via media, law and money, but despite their intentions and ambitions, the meaning of cultural landscape is always verified by everyday users.

Urban spaces carry a potential that hesitates between conformity and utopia, a world of commodities or of dreams. Today, urban places respond to market pressures, with public dreams defined by private development projects and public pleasures restricted to private entry. Two cultural products that most directly map the landscape are architecture and urban form. They both shape the city and our perception of it, but they are material symbols as well. Design and form relate to space in different ways: as a geographical constraint, as a terrain of potential conflict or cohesion, and as a commodity. Shopping centres have replaced political meetings and civic gatherings as arenas of public life. Despite private ownership and service to paying customers, they are perceived as a fairly democratic form of development. (Zukin 1993). The urban form has been especially vulnerable in recent years to an asymmetry of power favouring the private sector (see Figure 6.2).

Several questions have been raised concerning the symbolic places of recent past: how to re-interpret the objectionable history, what to remember, what to erase, what is important, and for whom? The discourse has been accelerated by various political goals and disputes. We cannot turn out our backs on the legacy of the past if we want to understand the present and plan the future. All of the old former communist icons are re-positioned and reinterpreted. Some buildings and monuments have often changed form, frequently function, and always significance. The red star is hardly anything more than a funky item, a party headquarters is not the political centre of the country, while a pair of blue jeans do not signify the desired West anymore. Many tracks of many *lieux mémoire* evoke different memories and connotations of socialist landscape icons. Those icons transfer us, just like Proust's *madeleine*, to the forgone landscapes, between reminiscence and oblivion; sometimes and for some of us this *madeleine* is sweet and tasty, but for some is only hard and bitter (Jaroszyńska and Jędrzejczak 2007). Hennelowa (2007, 3) advocates that 'the local population should have time and possibility to choose what they would like to erase from their memory, and what to leave and remember.' A 'down to top' approach seems to be most important and suitable, not only because it is the local people who have to live between relics of the past, and sometimes they choose to do it this way, but also because decisions declared from the top can be easily reversed, leaving local society confused.

Figure 6.2 Between old and new: central Bucharest, 2005

Cultural landscape is always a mode of communication or language. Since every language is a carrier of culture, the cultural landscape becomes a container and conveyer of multiple cultures. Histories and memories, expectations and aspirations, powers and weaknesses are reflected in constantly changing forms, functions and meanings of contemporary cities. The knowledge of society is coded in its practices and activities. Four decades of cynical and oppressive regimes left deep structural wounds in Central European societies, visualized as specific 'landscape scars'. Socially and economically emancipated societies have created emancipated landscapes based on the reinterpretation of old communist landscape features together with the implementation of new elements. It seems that still there are two parallel but interpenetrable landscapes: the old communist one and the new capitalist one. The main distinction between these two types is not based on form or function, but rather on a set of correspondences, connotations and even obsessions. This separation is significant only for older and more *tradition-oriented* person, determined by history and the past. It seems that the rebirth of nationalism in many countries of Central Europe is a response to the search for an all-explaining metanarration. The post-socialist categorization is much less important for the *other-directed* person, focused mainly on the future and knowing what he or she likes (see Riesman, Glazer and Denney 2001). For both types of people, cultural landscape remains an important carrier of cultural values. The emancipated post-communist landscape quite evidently reflects the objectives, abilities and contradictions of society. The patchwork landscape characterizes the urban life of the post-socialist cities. Glamorous and shining office and apartment towers often neighbour impoverished districts; gated communities accompany slums and ghettos.

The ideological landscape can be interpreted as visualization of 'happiness projects' (see Wagenaar 2004). But cultural landscape is also a moral value. Building a viable economy requires coherent moral values (Zukin 1993). The cultural landscape of Central Europe has undergone many transformations over the last 50 years: from bourgeois to Stalinist, then socialist modernist and since the early 1990s towards democratic, liberal and civic. The messages more or less clearly coded in the language of buildings and interpretations have changed from 'proletarian equivalence' towards 'civic discourse' but actually probably closer to 'arcadia of consumption'. Post-utopian landscapes of uncertainty and chaos also carry hopes for a better future. Some people search for new utopias, arcadias or more dogmatic 'happiness projects', while others are quite happy living in the landscape of plentiful choices and opportunities.

The cultural landscape of post-socialist cities can be seen as a living laboratory of transforming meanings and forms. The imprints of socialism that are left in the urban landscape are unusually deep and frequently affect the present outlook and structure of many cities. Central European cities carry the prints or stigmas of at least half a century of socialism. During the period of socialist supremacy, centrally steered circulation had been enforced in every urban and regional landscape. Interestingly, after 45 years of Moscow-dictated flows of styles and rules, the region jumped, often quite unconsciously, into another kind of circulation, dominated by unified and Western-style liberalism. The cultural landscapes have been quite rapidly transformed. The landscape of socialist triumphalism and egalitarian parity was inhabited and used by the masses, and usually designed and created despite their needs and expectations. Post-socialism brought many hopes and anticipations. The landscape is no longer facilitated by the Communist Party and local apparatchiks, but by new, still not fully realized, economic and civic powers. Societies are being transformed. On one hand there is growing public consciousness and awareness, and democratic practices negotiate functions and significances of cultural landscape features. But at the very same time, consumption facilitates many landscape projects. Citizens and consumers support landscape transformations towards a diverse and unknown future. The living landscape of supremacies and beliefs, of hopes and needs, of traumas and memories, creates a post-modern 'landscape blend', which to a great extent reveals our local, national and regional societies.

Bibliography

Adamik, M. (2001), 'The greatest promise – the greatest humiliation' in Jahnert, G. and Nickel, H.M. (eds), *Gender in transition* (Berlin: Trafo Verlag).

Ades, D., Benton, T., Elliott, D. and Whyte, I.B. (eds) (1995), *Art and power. Europe under the dictators 1930–1945* (London: Hayward Gallery).

Aitken, S. and Valentine, G. (eds) (2006), *Approaches to human geography* (London: Sage).

Ampelmann (2007), [homepage] http://www.ampelmann.de/ accessed on 20 September 2007.

Andrews, M. (2000), *Landscape and Western art* (Oxford: Oxford University Press).

Andrusz, G., Harloe, M. and Szelenyi, M. (eds) (1996), *Cities after socialism. Urban and regional conflict in post-socialist societies* (Oxford: Blackwell).

Argenbright, R. (1999), 'Remaking Moscow: new places, new selves' *Geographical review* 89(1), 1–22.

Arkadia (2007), [homepage] http://www.arkadia.com.pl/ accessed on 17 September 2007.

Armand, D.L. (1975), *Nauka o landshaftie* (Moskva: Mysl).

Ash, T.G. (1999), *History of the present* (New York: Random House).

Ashworth, G.J. (1998), 'The conserved European city as cultural symbol: the meaning of the text' in Graham, B. (ed.), *Modern Europe. Place. Culture. Identity* (London: Arnold).

Ashworth, G.J. (2002), 'Paradoksy i paradygmaty planowania przeszłości' in Purchla, J. (ed.), *Europa Środkowa. Nowy wymiar dziedzictwa.* (Kraków: Międzynarodowe Centrum Kultury).

Ashworth, G.J. and Tunbridge, J.E. (1999), 'Old cities, new pasts: heritage planning in selected cities of Central Europe', *GeoJournal* 49, 105–116.

Asiedu, D. (2005), 'World's biggest Stalin monument would have turned 50 on May Day' Radio Praha, 3 May 2005 [webpage] http://www.radio.cz/en/article/66095 accessed on 12 May 2007.

Atkins, P., Simmons, I. and Roberts, B. (1998), *People, land & time. An historical introduction to the relations between landscape, culture and environment* (London: Arnold).

Atkinson, D., Jackson, P., Sibley, D. and Washbourne, N. (eds) (2005), *Cultural geography. A critical dictionary of key concepts* (London: I.B. Tauris).

Aziz, H. (2001) 'Cultural keepers, cultural brokers: the landscape of woman and children – a case study of the Town Dahab in South Sinai' in Bender, B. and Winer M. (eds), *Contested landscapes. Movement, exile and place* (Oxford: Berg).

Baker, A.R.H. (2003), *Geography and history. Bridging the divide* (Cambridge: Press Syndicate of the University of Cambridge).

Barber, B. (2007), 'Czego chcesz chcieć', *Dziennik* 145(358), 17.

Basista, A. (2001), *Betonowe dziedzictwo. Architektura w Polsce czasów komunizmu* (Warszawa: Wydawnictwo Naukowe PWN).

Bauer, M. and Wicker, J. (2004), *Vorwärts immer. Rückwärts nimmer. 4000 Tage BRG* (Berlin: Nicolaische Verlagsbuchhandlung).

Bazylika Najświętszej Maryi Panny Licheńskiej (2007), [homepage] http://www.lichen.pl/index.php?t=page&dzial=1&sekcja=3 accessed on 17 September 2007.

Bender, B. (2001) 'Introduction' in Bender, B., Winer, M. (eds), *Contested landscapes. Movement, exile and place* (Oxford: Berg).

Bender, B. and Winer, M. (eds) (2001), *Contested landscapes. Movement, exile and place* (Oxford: Berg).

Ben-Joseph, E. (2005), *The code of the city. Standards and the hidden language of the place making* (Cambridge MA: MIT Press).

Benton, T. (1995), 'Speaking without adjectives. Architecture in the service of totalitarianism' in Ades D., Benton, T., Elliott, D. and Whyte, I.B. (eds), *Art and power. Europe under the dictators 1930–1945* (London: Hayward Gallery).

Berger, P.L. and Luckmann, T. (1966), *The social construction of reality. A treatise in the sociology of knowledge* (New York: Anchor Books).

Best, S. and Kellner, D. (1991), *Postmodern theory. Critical introductions* (New York: The Guilford Press).

Bielecki, Cz. (1996), *Gra w miasto* (Warszawa: Fundacja Dom Dostępny).

Bielecki, T. (2007), 'Estonia obala pomniki' *Gazeta Wyborcza* 13–14/01, 6.

Birks, H.H., Birks, H.J.B., Kaland, P.E., Moe, D. (eds) (2004), *The cultural landscape. Past, present and future* (Cambridge: Cambridge University Press).

Black, I.S. (2003), *(Re)reading architectural landscapes* in Robertson, I. and Richards, P. (eds) *Studying cultural landscapes* (London: Arnold), 19–46.

Blonsky, M. (1995), 'Introduction' in Blonsky, M. (ed.), *On signs* (Baltimore: Johns Hopkins University Press).

Blonsky, M. (ed.) (1995), *On signs* (Baltimore: Johns Hopkins University Press).

Bonnemaison, J. (2005), *Culture and space. Conceiving a new cultural geography* (London: I.B. Tauris).

Borén, Th. (2005), *Meeting places of transformation. Urban identity, spatial representations and local politics in St. Petersburg, Russia* (Stockholm: Department of Human Geography, Stockholm University).

Borley, L. (2002), 'Zobaczyć siebie jak widzą nas inni. Refleksje na temat tożsamości kulturowej' in Purchla, J. (ed.), *Europa Środkowa. Nowy wymiar dziedzictwa.* (Kraków: Międzynarodowe Centrum Kultury), 101–108.

Botton, A. de (2007), *The architecture of happiness. The secret art of furnishing your life* (London: Penguin).

Bourdieu, P. (1977), *Outline of a theory of practice* (Cambridge: Cambridge University Press).

Bourdieu, P., Thompson, J., Raymond, G. and Adamson, M. (1999), *Language and symbolic power* (Cambridge MA: Harvard University Press).

Brzezinski, Z.K. (1960), *The Soviet Bloc. Unity and conflict* (Cambridge MA: Harvard University Press).

Buttimer, A. (ed.) (1983), *The practice of geography* (London: Longman).

Buttimer, A. (2001), 'Sustainable development. Issues of scale and appropriateness' in Buttimer, A. (ed.), *Sustainable landscapes and lifeways* (Cork: Cork University Press), 7–34.

Buttimer, A. (ed.) (2001), *Sustainable landscapes and lifeways* (Cork: Cork University Press).

Calvino, I. (1978), *Invisible cities* (San Diego: Harcourt).

Carravetta, P. (1998), 'The reasons of the code. Reading Eco's A theory of semiotics' in Silverman, H.J. (ed.), *Cultural semiosis. Tracing the signifier* (London: Routledge).

Casey, E.S. (1987), *Remembering. A phenomenological study* (Bloomington: Indiana University Press).

Centrul Civic (2007), [homepage] http://en.wikipedia.org/wiki/Centrul_Civic accessed on 17 September 2007.

Certeau, M. de (1985), *Practices of space* in Blonski, M. (ed.), *On signs* (Baltimore: Johns Hopkins University Press), 122–145.

Chomsky, N. (2004), *Hegemony or survival. America's quest for global dominance (American Empire Project)* (New York: Henry Holt).

Clark, K. (1956), *Landscape into art* (Harmondsworth: Penguin).

Cohen, J.-L. (1995), 'When Stalin meets Haussmann. The Moscow Plan of 1935' in Ades, D., Benton, T., Elliott, D. and Whyte, I. B. (eds), *Art and power. Europe under the dictators 1930–1945* (London: Hayward Gallery).

Communism (2007), [homepage] Encyclopædia Britannica Online http://www. britannica.com/eb/article-9117284/communism accessed on 13 June 2007.

Cook, I., Crouch, D., Naylor, S. and Ryan, J.R. (eds) (2000), *Cultural turns/ geographical turns. Perspectives on cultural geography* (Harlow: Prentice Hall).

Cook, I., et al. (2005), Positioning/situated knowledge in Atkinson, D., Jackson P., Sibley D. and Washbourne, N. (eds) *Cultural geography. A critical dictionary of key concepts* (London: I.B. Tauris).

Corner, J. (2006), *Terra fluxus* in Waldheim, Ch. (ed.), *The Landscape Urbanism Reader* (New York: Princeton Architectural Press), 23–33.

Cosgrove, D. and Jackson, P. (1987), 'New directions in cultural geography' *Area*, 19, 95–101.

Cosgrove, D.E. (1984), *Social formation and symbolic landscape* (London: Croom Helm).

Cosgrove, D.E. (1993), *The Palladian landscape. Geographical change and its cultural representations in sixteen-century Italy* (Leicester: Leicester University Press).

Cosgrove, D.E. and Daniels, S. (eds) (2004), *The iconography of landscape. Essays on the symbolic representation, design and use of past environments* (Cambridge: Cambridge University Press).

Crampton, R. and Crampton, B. (2002), *Atlas of Eastern Europe in the twentieth century* (London: Routledge).

Crang, M. (2004), *Cultural geography* (London: Routledge).

Crang, M. and Thrift, N. (eds) (2002), *Thinking space* (London: Routledge).

Cresswell, T. (2004), *Place. A short introduction* (Malden: Blackwell).

Crowley, D. and Reid, S.E. (eds) (2002), *Socialist spaces. Sites of everyday life in the Eastern Bloc* (Oxford: Berg).

Cupers, K. and Miessen, M. (2002), *Spaces of uncertainty* (Wuppertal: Müller + Busmann).

Czepczyński, M. (2004), 'Living the cultural landscape. The heritage and identity of the inhabitants of Gdański' in Sagan, I. and Czepczyński, M. (eds.) (2004), *Featuring the quality of urban life in contemporary cities of Eastern and Western Europe* (Gdańsk: Katedra Geografii Ekonomicznej Uniwersytetu Gdańskiego – Bogucki Wydawnictwo Naukowe).

Czepczyński, M. (2005a), 'Post-socialist landscape management and reinterpretation' in Oliver, L., Millar, K., Grimski, D., Ferber, U. and Nathanail, C.P. (eds), *CABERNET 2005. Proceedings of CABERNET 2005. The international conference on managing urban land* (Nottingham: Land Quality Press).

Czepczyński, M. (2005b), 'De- vs. re-industrialisation of post-socialist city. The case of Gdańsk' *Regions* 259, 6–11.

Czepczyński, M. (2006a), 'Transformations of Central European cultural landscapes. Between circulations and iconography'. *Bulletin of Geography. Socio-Economic Series* 6, 5–15.

Czepczyński, M. (2006b), 'Understanding cultural landscapes. Approaches and practices'. *Past Place. Newsletter of the Historical Geography Specialty Group. The Association of American Geographers* 15(1), 5–6.

Czepczyński, M. (2006c), 'Krajobraz kulturowy miast po socjalizmie. Tendencje przemian form i znaczeń' in Czepczyński, M. (ed.), *Przestrzenie miast post-socjalistycznych. Studia społecznych przemian przestrzeni zurbanizowanej* (Gdańsk: Katedra Geografii Ekonomicznej Uniwersytetu Gdańskiego – Bogucki Wydawnictwo Naukowe).

Czepczyński, M. (ed.) (2006d), *Przestrzenie miast post-socjalistycznych. Studia społecznych przemian przestrzeni zurbanizowanej* (Gdańsk: Katedra Geografii Ekonomicznej Uniwersytetu Gdańskiego – Bogucki Wydawnictwo Naukowe).

Czepczyński, M. (2007), 'Podejścia badawcze w nowej geografii kultury' in Maik, W., Rembowska, K. and Suliborski, A. (eds), *Geografia a przemiany współczesnego świata* (Bydgoszcz: Wydawnictwo Uczelniane WSG).

Dahbour, O. and Ishay M.R. (eds) (1995), *The nationalism reader* (New York: Humanity Books).

Darby, W.J. (2000), *Landscape and identity. Geographies of nation and class in England* (Oxford: Berg).

David, P.A. (2001), 'Path dependence, its critics and the quest for 'historical economics' in Garrouste, P. and Ioannides, S. (eds), *Evolution and path dependence in economic ideas. Past and present* (Cheltenham: Edward Elgar).

Davis, M. (1990), *City of quartz. Excavating the future in Los Angeles* (London: Verso).

DDR Museum (2007) [homepage] www.ddr-museum.de accessed on 13 May 2007.

Dear, M.J. and Flusty, S. (eds) (2002), *The spaces of postmodernity. Readings in human geography* (Oxford: Blackwell).

Denzer, V. (2005), *On the (re-) construction of urban histories in times of transition.* Manuscript (Leipzig: Universität Leipzig, Institut für Geographie).

Domański, B. (1997), *Industrial control over the socialist town. Benevolence or exploitation?* (Westport: Praeger).

Domański, B. (2004a), 'Moral problems of Eastern wilderness: European core and periphery' in Lee, R. and Smith, D.M. (eds), *Geographies and moralities. International perspectives on development, justice and place* (Oxford: Blackwell).

Domański, B. (2004b), 'West and east in 'new Europe'. The pitfalls of paternalism and a claimant attitude' *European Urban and Regional Studies* 11, 377–381.

Domański, H. (2002), *Ubóstwo w społeczeństwach postkomunistycznych* (Warszawa: Instytut Spraw Publicznych).

Dorrian, M. and Rose, G. (2003), 'Introduction' in Dorrian, M. and Rose, G., *Deterritorialisations... Revisioning. Landscape and politics* (London: Black Dog), 13–19.

Dovey, K. (2001), *Framing places. Mediating power in built form* (London: Routledge).

Dreichfuß, H. (1987), *Rumänische Rhapsodie* (Berlin: Verlag der Nation).

Driver, F. and Gilbert, D. (eds) (1999), *Imperial cities. Landscape, display and identity* (Manchester: Manchester University Press).

Dudek, A. (2005), *Śady PeeReLu. Ludzie, Wydarzenia, Mechanizmy* (Kraków: Arcana/Instytut Historii PAN).

Dunauvaros (2007), [webpage] http://www.geocities.com/Eureka/Plaza/7807/dunaujvaros/ accessed on 18 October 2007.

Duncan, J.S. (2004), *City as text, the politics of landscape interpretation in the Kandyan Kingdom* (Cambridge: Cambridge University Press).

Dutkiewicz, W. (2005), *Ślady po PeeReLu. Nieodległa rzeczywistość* (Warszawa: QLCO).

Dvořák, K. (1982), *Zásady a kritéria ideovosti architektury* (Prague: Vúva).

Eco, U. (1985), 'Producing signs' in Blonsky, M. (ed.), *On signs* (Baltimore: Johns Hopkins University Press).

Eco, U. (1986), *Semiotics and the philosophy of language* (Bloomington: Indiana University Press).

Eisenhuetenstadt (2007), [homepage] http://www.eisenhuettenstadt.de/cgi-bin/mainen.php?d1cnr=76 accessed on 18 October 2007.

Enyedi, G. (1996), 'Urbanisation under socialism' in Andrusz, G., Harloe, M. and Szelenyi, M. (eds), *Cities after socialism. Urban and regional conflict in post-socialist societies* (Oxford: Blackwell).

Enyedi, G. (1998), 'Transformation in Central European postsocialist cities' in Enyedi, G. (ed.), *Social change and urban restructuring in Central Europe* (Budapest: Akadémiai Kiadó).

Enyedi, G. (ed.) (1998), *Social change and urban restructuring in Central Europe* (Budapest: Akadémiai Kiadó).

Esterházy, P. (2007), 'Donosy zawsze szkodzą' *Dziennik – Europa* 23(166), 14–16.

European Landscape Convention (2000), [website] Council of Europe, http://www.coe.int/t/e/Cultural_Co-operation/Environment/Landscape/ accessed on 30 May 2007.

Fægri, K. (2004) 'Preface' in Birks, H.H., Birks, H.J.B., Kaland P.E. and Moe, D. (eds), *The cultural landscape. Past, present and future* (Cambridge: Cambridge University Press), 1–4.

Fodor, J.A. (1981), *Representations. Philosophical essays on the foundations of cognitive science* (Cambridge MA: MIT Press).

Foote, K., Tóth, S. and Arvay, A. (2000), 'Hungary after 1989. Inscribing a new past on place' *Geographical Review* 90(3), 301–334.

Forman, R.T.T. (1995), *Land mosaics. The ecology of landscapes and regions* (Cambridge: Cambridge University Press).

Fortuna, G. and Tusk, D. (1990), *Wydarzyło się w Gdańsku 1901 – 2000. Jeden wiek w jednym mieście* (Gdańsk: Millenium Media).

Foucault, M. (1975), *Discipline and punish. The birth of the prison* (New York: Random House).

Foucault, M. (1986), 'Space, knowledge, and power.' in Rabinow, P. (ed.), *The Foucault reader. An introduction to Foucault's thought* (London: Harmondsworth).

Foucault, M. and Gordon, C. (1981), *Power/Knowledge. Selected interviews and other writings 1972–1997 by Michel Foucault* (New York: Pantheon Books).

Fowkes, R. (2002), 'The role of monumental sculptures in the construction of socialist space in Stalinist Hungary' in Crowley, D. and Reid, S.E. (eds) (2002), *Socialist spaces. Sites of everyday life in the Eastern Bloc* (Oxford: Berg), 65–84.

Gądecki, J. (2005), *Architektura i tożsamość. Rzecz o antropologii architektury* (Nowa Wieś: Wydawnictwo Rolewski).

Garrouste, P. and Ioannides, S. (eds) (2001), *Evolution and path dependence in economic ideas. Past and present* (Cheltenham: Edward Elgar).

Geertz, C. (1973), *The interpretation of culture* (New York: Crossroad).

Gennep, A. van (1960), *The rites of passage* (Chicago: Chicago University Press).

Ghirardo, D. (1999), *Architektura po modernizmie* (Toruń: VIA).

Giddens, A. (ed.) (1974), *Positivism and sociology* (London: Heinemann).

Gierasimov, I.P. (1966), *Kanstruktivnaya gieografia: ciely, mietody. rezultaty* (Moskva: Izviestia VFO).

Goldzamt, E. and Szwidkowski, O. (1987), *Kultura urbanistyczna krajów socjalistycznych. Doświadczenia europejskie* (Warszawa: Arkady; Moskva: Strojizdat).

Gottdiener, M. (1997), *The social production of urban space* (Austin: University of Texas Press).

Gottmann, J. and Harper, R.A. (1990), *Since Megalopolis. The urban writings of Jean Gottmann* (Baltimore: John Hopkins University Press).

Graham, B. (1998), 'The past in Europe's present. Diversity, identity and the construction of place' in Graham, B. (ed.), *Modern Europe. Place. Culture. Identity* (London: Arnold).

Graham, B. (ed.) (1998), *Modern Europe. Place. Culture. Identity* (London: Arnold).

Graham, B. and Ashworth, G.J. (2000), *Geography of heritage. Power, culture and economy* (London: Arnold).

Gray, J. (2007), *Black masses. Apocalyptic religion and death of Utopia* (London: Allen Lane).

Green, P.S. (1998), '2 Czech cities to wall off their "problematic" Gypsies' *International Herald Tribune*, 25 May [webpage] http://www.iht.com/articles/1998/05/25/gypsy.t.php accessed on 13 June 2007.

Groys, B. (1988), *Gesamtkunstwerk Stalin. Die gespaltene Kultur in der Sowjetunion* (München: Hanser).

Guillemoles, A. (2007), Le Populism se leve a l'Est *Politique Internationale – La Revue* 114, http://www.politiqueinternationale.com/revue/read2.php?id_revue=1 14&id=608&content=texte accessed on 10 August 2007.

Habermas, J. (1984), *The theory of communicative action, volume 1. Reason and the rationalization of society* (Boston: Beacon Press).

Hacking, I. (1999), *The social construction of what?* (Cambridge MA: Harvard University Press).

Halbwachs, M. and Coser, L.A. (1992), *On collective memory* (Chicago: University of Chicago Press).

Hale, J.A. (2000), *Building ideas. An introduction to architectural theory* (Chichester: John Wiley & Sons).

Hall, S. (2002), 'The work of representation' in Hall, S. (ed.), *Representation. Cultural representation and signifying practices* (London: Sage).

Hall, S. (ed.) (2002), *Representation. Cultural representation and signifying practices* (London: Sage).

Hanley, S. (1999), 'Concrete conclusions. The discreet charm of the Czech panelak' *Central Europe Review* 22, http://www.ce-review.org/authorarchives/hanley_archive/hanley22old.html accessed on 13 June 2007.

Hannemann, Ch., Kabisch, S. and Weiske, Ch. (eds) (2002), *Neue Länder – Neue Sitten? Transformationsprozesse in Städten und Regionen Ostdeutschlands* (Berlin: Schiler).

Harding, L. (2007), 'Protest by Kremlin as police quell riots in Estonia' *Guardian Unlimited*, 29 April 2007 [webpage] http://www.guardian.co.uk/russia/article/0,,2068215,00.html accessed on 12 May 2007.

Harloe, M. (1996), 'Cities in the Transition' in Andrusz, G., Harloe, M. and Szelenyi, M. (eds.), *Cities after socialism. Urban and regional conflict in post-socialist societies* (Oxford: Blackwell).

Harrison, P. (2006), 'Post-structuralist theories' in Aitken, S. and Valentine, G. (eds), *Approaches to human geography* (London: Sage).

Harvey, D. (1982), *The limits to capital* (Oxford: Blackwell).

Hauben, Th. (2004), 'Places without a sense of place. New icons in Central and Eastern Europe' in Wagenaar, C. and Dings, M. (eds.), *Ideals in concrete. Exploring Central and Eastern Europe* (Rotterdam: NAi Publishers).

Hayek, F.A. von (1988). 'The fatal conceit. The errors of socialism' in Bartley, W.W., III (ed.), *The Collected Works of Friedrich August Hayek*, vol. I (London: Routledge).

Heidegger, M., Fried, G. and Polt, R. (2001), *Introduction to Metaphysics* (New Haven: Yale University Press – Yale Nota Bene).

Hennelowa, J. (2007), 'Lekcja Tallina' *Tygodnik Powszechny*, 3.

Hertle, H.-H. and Wolle, S. (2002), *Damals in der DDR. Der Alltag im Arbeiter- und Bauernstadt* (München: C. Bertelsmann).

Hilkovitch, J. and Fulkerson, M. (1997), *Paul Vidal de la Blache. A biographical sketch.* [webpage] http://wwwstage.valpo.edu/geomet/histphil/test/vidal.html updated 25 March 1997, accessed 20 September 2007.

Hirsch, E. (2003), 'Introduction' in Hirsch, E. and O'Hanlon, M. (eds), *The anthropology of landscape. Perspectives on place and space* (Oxford: Oxford University Press).

Hirsch, E. and O'Hanlon, M. (eds) (1995), *The anthropology of landscape. Perspectives on place and space* (Oxford: Oxford University Press).

Hirszowicz, M. (1980), *The bureaucratic Leviathan. A study in the sociology of communism* (Oxford: Martin Robertson).

History of PKiN in a Nutshell (2007), [homepage] http://www.pkin.pl/historia/ accessed on 18 June 2007.

Hitchcock, W.I. (2004), *The struggle for Europe. The history of the continent since 1945* (London: Profile Books).

Hörschelmann, K. (2002), 'History after the end. Post-socialist difference in a (post)modern world' *Transactions of the Institute of British Geographers* 27, 52–66.

Ingegnoli, V. (2004), *Landscape ecology. A widening foundation* (New York: Springer-Verlag).

Institute for the Investigation of Communist Crimes in Romania (2006), [homepage] http://www.crimelecomunismului.ro/en/about_iiccr accessed on 30 May 2007.

Ioan, A. (1999), 'A postmodern critics' kit for interpreting socialist realism' in Leach, N. (ed.), *Architecture and revolution. Contemporary perspectives on Central and Eastern Europe* (London: Routledge).

Ioan, A. (2007), 'The peculiar history of (post)communist public places and spaces: Bucharest as a case study' in Stanilov, K. (ed.), *The post-socialist city. Urban form and space transformations in Central and Eastern Europe* (Dordrecht: Springer).

Jackson, J.B. (1984), '*The world itself*'. *Discovering vernacular landscape* (New Haven: Yale University Press).

Jahnert, G. and Nickel, H.M. (eds) (2001), *Gender in transition* (Berlin: Trafo Verlag).

Jakubowska, U. and Skarżyńska, K. (eds) (2005), *Demokracja w Polsce: Doświadczanie zmiany* (Warszawa: Wydawnictwa Instytutu Psychologii PAN).

Jaroszyńska, K. and Jędrzejczak, A. (2007), 'Za co kochamy PRL' *Przekrój*, 29/3239, 31–35.

Jay, M. (1998), *Cultural semantics. Keywords of our time* (Amherst: University of Massachusetts Press).

Jencks, Ch. (2005), *The iconic building. The power of enigma* (London: Frances Lincoln).

Johnson, R., Chambers, D., Raghuram, P. and Tincknell, E. (2004), *The practice of cultural studies* (London: Sage).

Judt, T. (2005), *Postwar. A history of Europe since 1945* (New York: The Penguin Press).

Kaluza, T. (2007), *The promotion of cooperation between Romas and local government through interactive seminars* [webpage] http://lgi.osi.hu/ethnic/csdb/results.asp?idx=no&id=90 accessed on 17 August 2007.

Kapuściński, R. (1982), *Szachinszach* (Warszawa: Czytelnik).

Kapuściński, R. (2007), *Lapidarium IV* (Warszawa: Czytelnik).

Kazepov, Y. (2005), 'Cities of Europe. Changing contexts, local arrangements, and the challenge to social cohesion' in Kazepov, Y. (ed.), *Cities of Europe, changing contexts, local arrangements, and the challenge to urban cohesion* (Malden: Blackwell).

Kazepov, Y. (ed.) (2005), *Cities of Europe, changing contexts, local arrangements, and the challenge to urban cohesion* (Malden: Blackwell).

Kieniewicz, J. (2002), 'Stojąc w drzwiach: odczytywanie dziedzictwa i wybór przynależności' in Purchla, J. (ed.), *Europa Środkowa. Nowy wymiar dziedzictwa* (Kraków: Międzynarodowe Centrum Kultury).

Koch, A.M. (2007), *Poststructuralism and the politics of method* (Lanham: Lexington Books).

Kołodziejczyk, M. (2007), 'Zabytkowicze i awangardziści. Gruzy Warszawy jako wyzwolenie', *Polityka. Pomocnik Historyczny* 34(2617), 22–29.

Kong, L.L.L. (2007), *A 'new' cultural geography? Debates about invention and reinvention* [webpage] http://profile.nus.edu.sg/fass/geokongl/scotgeom.pdf accessed 21 March.

Kopleck, M. (2006), *Berlin 1945–1989. Past finder* (Berlin: Ch. Links).

Kotarbiński, A. (1967), *Rozwój architektury i urbanistyki polskiej w latach 1944–1964* (Warszawa: Arkady).

Kovács, Z. and Wiessner, R. (eds) (1997), Prozesse und Perspektiven der Stadtentwicklung in Ostmitteleuropa, *Münchener Geographische Hefte* 76, Geographischen Institut der Technischen Universität München (Passau: L.I.S. Verlag).

Kremzerová, D. (2006), *Malý Przewodnik po Dużej Ostrawie* (Ostrava: Ostravský informační serwis).

Krez, J. (1997), *Architektura znaczeń* (Gdańsk: Politechnika Gdańska).

Król, K. (2007), 'Jak porzuciłem skrzela' *Duży Format. Gazeta Wyborcza* 34(745), 10–12.

Kurczewski, J. (2007), 'Młodzi chcą galopować *Polityka*', 4, 27 January, 30–31.

Laclau, E. (1990), *New reflections on the revolution of our time* (London: Verso).

Largo Sofia (2007), [homepage] http://en.wikipedia.org/wiki/Largo%2C_Sofia accessed on 17 September 2007.

Lawson, B. (2003), *The language of space* (Amsterdam: Architectural Press).

Leach, N. (1999), 'Architecture or Revolution?' in Leach, N. (ed.), *Architecture and revolution. Contemporary perspectives on Central and Eastern Europe* (London: Routledge).

Leach, N. (ed.) (1999), *Architecture and revolution. Contemporary perspectives on Central and Eastern Europe* (London: Routledge).

Lee, R. and Smith, D.M. (eds) (2004), *Geographies and moralities. International perspectives on development, justice and place* (Oxford: Blackwell).

Lefebvre, H. (1991), *The production of space* (Oxford: Blackwell).

196 *Cultural Landscapes of Post-Socialist Cities*

LeGates, R. and Stout F. (eds) (2000), *The city reader* (London: Routledge).

Lévinas, E. (2003), *Humanism of the other* (Chicago: University of Illinois Press).

Lewicka, M. (2005), 'Architektura spoleczeństwa obywatelskiego. O roli miejsca zamieszkania w budowaniu tożsamości lokalnej' in Jakubowska, U. and Skarżyńska, K. (eds), *Demokracja w Polsce: Doświadczanie zmiany* (Warszawa: Wydawnictwa Instytutu Psychologii PAN).

Lewicka, M. and Zaborska, K. (2007), ' Osiedla zamknięte – czy istnieje alternatywa?' *Kolokwia Psychologiczne*, 16, 135–152.

Lewis, Ch. P. (2005), *How the East was won. Impact of multinational companies in Eastern Europe and the former Soviet Union* (Houndmills: Palgrave Macmillan).

Lovenduski, J. and Woodall, J. (1987), *Politics and society in Eastern Europe* (Bloomington: Indiana University Press).

Low, S. (2001), 'The edge and the center. Gated communities and the discourse of urban fear' *American Anthropologist* 103(1), 45–58.

Łukasiewicz, J. (1996) 'Mitologie socrealizmu' *Odra* 11, 36–42.

Lynch, K. (1960), *The Image of the City* (Cambridge MA: MIT Press).

Mach, Z. (2006), 'Multicultural heritage, remembering, forgetting, and the construction of identity' in Schröder-Esch, S. and Ulbricht, J.H. (eds), *The politics of heritage and regional development strategies – actors, interests, conflicts* (Weimar: Bauhaus Universität).

Maik, W., Rembowska, K. and Suliborski, A. (eds) (2007), *Geografia a przemiany współczesnego świata* (Bydgoszcz: Wydawnictwo Uczelniane WSG).

Marcus, G.E. (2000), 'The twistings and turnings of geography and anthropology in winds of millennial transition' in Cook, I., Crouch, D., Naylor, S. and Ryan, J.R. (eds), *Cultural turns/geographical turns. Perspectives on cultural geography* (Harlow: Prentice Hall), 14–25.

Marina Mokotow (2007), [homepage] http://www.marinamokotow.com.pl/english/content.html accessed on 20 October 2007.

Markus, T.A. and Cameron, D. (2002), *The words between the spaces. Buildings and landscapes* (London: Routledge).

Martin, G.J. and James, P.E. (1993), *All possible worlds. A history of geographical ideas* (New York: John Wiley and Sons).

Marx, K. and Engels, F. (2002), *The Communist Manifesto* (London: Penguin).

Massey, D. (2006) *For space* (London: Sage).

Meining, D.W. (1979), 'Introduction' in Meining, D.W. (ed.), *Interpretation of ordinary landscapes, geographical essays* (Oxford: Oxford University Press).

Meining, D.W. (ed.) (1979), *Interpretation of ordinary landscapes, geographical essays* (Oxford: Oxford University Press).

Memorial of the Victims of Communism (2007), [homepage] http://www.memorialsighet.ro/en/istoric_cladire_sighet.asp accessed on 20 September 2007.

Merrifield, A. (2002), 'Henri Lefebvre. A socialist in space' in Crang, M. and Thrift, N. (eds), *Thinking space* (London: Routledge).

Mezga, D. (1998) 'Political Factors in Reconstruction of Warsaw's Old Town' *Urban Design Studies* 4, 16–28.

Minkey, G. and Rassool, C. (1998) 'Orality, Memory, and Social History in South Africa' in Nuttall, S. and Coetzee, C. (eds), *Negotiating the past. The making of memory in South Africa* (Cape Town: Oxford University Press).

Mitchell, D. (2001), *Cultural geography. A critical introduction* (Oxford: Blackwell).

Mitchell, D. (2005), 'Landscape' in Atkinson, D., Jackson, P., Sibley, D. and Washbourne, N. (eds), *Cultural geography. A critical dictionary of key concepts* (London: I.B. Tauris), 49–56.

Mitchell, W.J. (2005), *Placing words. Symbols, space, and the city* (Cambridge MA: MIT Press).

Modrzejewski, F. and Sznajdeman, M. (eds) (2002), *Nostalgia. Eseje o tęsknocie za komunizmem* (Wołowiec: Czarne).

Murawski, J. (2007), 'Ucieczka na wieże' *Polityka* 39(2622), 94–97.

Museum of Communism (2007), [homepage] http://www.muzeumkomunismu.cz/ accessed on 20 September 2007.

Naveh, Z. and Lieberman, A. (1984), *Landscape ecology. Theory and application* (New York: Springer-Verlag).

Nawratek, K. (2005), *Ideologie w przestrzeni. Próby demistyfikacji* (Kraków: Universitas).

Neelen, M. and Dzokic, A. (2004), 'Forget about design' in Wagenaar, C. and Dings, M. (eds), *Ideals in concrete. Exploring Central and Eastern Europe* (Rotterdam: NAi Publishers).

Neill, W.J.V. (2004), *Urban planning and cultural identity* (London: Routledge).

Niechciane dziedzictwo. Różne oblicza architektury nowoczesnej w Gdańsku i Sopocie. Unwanted Heritage. Various Faces of the Architectural Modernity in Gdańsk and Sopot (2005) (Gdańsk: Centrum Sztuki Współczesnej Łaźnia).

Nora, P. (2006), *Rethinking France. Les Lieux de memoire, Volume 2: Space* (Chicago: University of Chicago Press).

Norberg–Schulz, Ch. (1971), *Existence, space and architecture* (London: Studio Vista).

Norberg–Schulz, Ch. (1999), *Znaczenie w architekturze Zachodu*, Wyd. (Warszawa: Murator).

Nowa Huta (2007), [homepage] http://www.nh.pl/ accessed on 10 August 2007.

Nowak, A. (2004), *Od imperium do imperium. Spojrzenia na historię Europy Wschodniej* (Kraków: Arcana/Instytut Historii PAN).

Nuttall, S. and Coetzee, C. (eds) (1998) *Negotiating the past. The making of memory in South Africa* (Cape Town: Oxford University Press).

Oliver, L., Millar, K., Grimski, D., Ferber, U. and Nathanail, C.P. (eds) (2005), *CABERNET 2005. Proceedings of CABERNET 2005: The International Conference on Managing Urban Land.* (Nottingham: Land Quality Press).

Olszański, T. (1970), *Budapeszteńskie ABC* (Warszawa: Iskry).

Olzacki, S. (2007), 'Przerwany Kongres Kultury', *Polityka* 50(2590), 100–105.

Ortega y Gasset, J. (1994), *The revolt of the masses* (New York: W.W. Norton & Co.).

Ortega y Gasset, J. (1996), *Man and people* (New York: W.W. Norton & Co.).

Orwell, G. (1949), *Nineteen eighty-four. A novel* (London: Secker & Warburg).

Oswalt, Ph. (ed.) (2006), *Shrinking cities. Vol. 1. International research* (Ostfildern-Ruit: Hatje Cantz).

Paci, P., Sasin, M.J. and Verbeek, J. (2004), *Economic Growth, Income Distribution and Poverty in Poland during Transition* [webpage] http://www-wds.worldbank.org/servlet/WDSContentServer/WDSP/IB/2005/01/19/000160016_20050119122932/Rendered/PDF/WPS3467.pdf accessed on 27 April 2007.

Painter, J. and Philo, C. (1995) 'Spaces of citizenship. An introduction' *Political geography* 14(2).

Palace of the Parliament (2007), http://en.wikipedia.org/wiki/Palace_of_the_Parliament accessed on 17 September 2007.

Palang, H., Sooväli, H. and Antrop, M. (eds) (2004), *European rural landscapes. Persistence and change in a globalizing environment* (Boston: Kluwer Academic Publishers).

Panofsky, E. (1972), *Studies in iconology. Humanistic themes in the art of the Renaissance* (New York: Harper & Row).

Papanek, V. (1997), *Design for the real world. Human ecology and social change* (London: Thames and Hudson).

Paszkowiak, A. and Pelzer, H. (1976), *Bulgarien* (Leipzig: VEB F.A. Brockhaus).

Petrescu, D. (1999), 'The People's House, or the voluptuous violence of an architectural paradox' in Leach, N. (ed.), *Architecture and revolution. Contemporary perspectives on Central and Eastern Europe* (London: Routledge).

Philo, Ch. (2000), *More words, more worlds. Reflections of the 'cultural turn' and human geography* in Cook, I., Crouch, D., Naylor, S. and Ryan, J.R. (eds), *Cultural turns/geographical turns. Perspectives on cultural geography* (Harlow: Prentice Hall).

Piaget, J. (1968), *The psychology of intelligence* (New York: Littlefield).

Pietrasik, Z. (2007), *PRL do śmiechu*. Polityka 29(2613), 59–61.

Pinker, S. (1991), 'Rules of language' *Science* 253, 530–535.

Piński, A. and Piński, J. (2007), 'Unia głupich inwestycji' *Wprost* 39, 66–69.

Piwowarczyk, M. (2005), 'Profit and Luxury' *The Warsaw Voice* 12.10.

Przytułek dla pomników PRL, 2007. Polityka 23(2607), 17.

Purchla, J. (ed.) (2002), *Europa Środkowa. Nowy wymiar dziedzictwa* (Kraków: Międzynarodowe Centrum Kultury).

Raabe, K. and Sznajderman, M. (eds.) (2006), *Znikająca Europa* (Czarne: Wołowiec).

Rabinow, P. (ed.) (1986), *The Foucault reader. An introduction to Foucault's thought* (London: Harmondsworth).

Reid, S.E. and Crowley D. (eds) (2002), *Style and socialism. Modernity and material culture in post-war Eastern Europe* (Oxford: Berg).

Rembowska, K. (1998), 'Przestrzeń znacząca. Miasto socjalizmu' *Kwartalnik Geograficzny*, 4(8), 11–14.

Removal of the Palast der Republik (2007), [homepage] http://www.stadtentwicklung.berlin.de/bauen/index_en.shtml accessed on 17 September 2007.

Renan, E. (1995), 'What is nation' in Dahbour, O. and Ishay, M.R. (eds), *The nationalism reader* (New York: Humanity Books).

Repa, J. (2000), 'Czech court backs anti-Gypsy wall' *BBC News*, April 12, [webpage] http://news.bbc.co.uk/2/hi/europe/711211.stm accessed on 13 June 2000.

Revolutions of 1989 (2007) [webpage] http://en.wikipedia.org/wiki/Revolutions_ of_1989 accessed on 17 September 2007.

Ricoeur, P. (2004), *Memory, history, forgetting* (Chicago: University of Chicago Press).

Riesman, D., Glazer, N. and Denney, R. (2001), *The lonely crowd, revised edition. A study of the changing American character* (New Haven: Yale Nota Books).

Ritzer, G. (2007), *Globalization of nothing* (Thousand Oaks: Sage).

Robertson, I. and Richards, P. (eds) (2003), *Studying cultural landscapes* (London: Arnold).

Ryan, J.R. (2000), 'Introduction' in Cook, I., Crouch, D., Naylor, S. and Ryan, J.R. (eds) (2000), *Cultural turns/geographical turns. Perspectives on cultural geography* (Harlow: Prentice Hall), 10–12.

Rychling, A. and Solon, J. (1998), *Ekologia krajobrazu* (Warszawa: PWN).

Rykwert, J. (2000), *The seduction of place. The history and future of the city* (Oxford: Oxford University Press).

Sagan, I. and Czepczyński, M. (eds) (2004), *Featuring the quality of urban life in contemporary cities of Eastern and Western Europe* (Gdańsk: Katedra Geografii Ekonomicznej Uniwersytetu Gdańskiego – Bogucki Wydawnictwo Naukowe).

Said, E. (2000), 'Invention, memory, and place' *Critical inquiry* 26(2), 175–192.

Salvadori, R. (2004), *Mitologia nowoczesności* (Warszawa: Zeszyty Literackie).

Sármány-Parsons, I. (1998), 'Aesthetic aspects of change in urban space in Prague and Budapest during the transition' in Enyedi, G. (ed.), *Social change and urban restructuring in Central Europe* (Budapest: Akadémiai Kiadó).

Satjukow, S. and Gries, R. (eds) (2002), *Soziliastische Helden. Eine Kulturgeschichte von Propogandafiguren in Osteuropa und der DDR* (Belrin: Ch. Links).

Sauer, C.O. (1925), 'The morphology of landscape', *University of California publications in geography* 2, 19–54.

Saussure, F. de (1974), *Course in general linguistics* (London: Fontana).

Schama, S. (1996), *Landscape and memory* (New York: Alfred A. Knopf).

Schröder-Esch, S. (ed.) (2006), *Practical aspects of cultural heritage – presentation, revaluation, development* (Weimar: Bauhaus Universität).

Schröder-Esch, S. and Ulbricht, J.H. (eds) (2006), *The politics of heritage and regional development strategies – actors, interests, conflicts* (Weimar: Bauhaus Universität).

Sennett, R. (1990), *The conscience of the eye. The design and social life of cities* (New York: W.W. Norton & Company).

Shanker, T. and Mazzetti, M. (2007), 'New defense chief eases relations Rumsfeld bruised' *New York Times* [webpage] 12 March 2007 http://query.nytimes.com/ gst/fullpage.html?res=9E0CE1DB1031F931A25750C0A9619C8B63 accessed on 15 May 2007.

Shannon, K. (2006), 'From theory to resistance: landscape urbanism in Europe' in Waldheim, Ch. (ed.), *The landscape urbanism reader* (New York: Princeton Architectural Press).

Shaw, D.J.B. and Oldfield, J.D. (2007), 'Landscape science. A Russian geographical tradition' *Annals of the Association of American Geographers* 91(1), 111–126.

Shurmer-Smith, P. (ed.) (2002), *Doing cultural geography* (London: Sage).

Shurmer-Smith, P. and Hannam, K. (1994), *Worlds of desire, realms of power. A cultural geography* (London: Verso).

Silverman, H.J. (ed.) (1998), *Cultural semiosis. Tracing the signifier* (London: Routledge).

Šimečka, M. (2002), '110 konarów. Realny socjalizm i płynące zeń nauki' in Modrzejewski, F. and Sznajdeman, M. (eds), *Nostalgia. Eseje o tęsknocie za komunizmem* (Wołowiec: Czarne).

Simons, T.W., Jr. (1993), *Eastern Europe in the Postwar World* (Houndmills: Macmillian).

Smith, D.M. (1996), 'The socialist city' in Andrusz, G., Harloe, M. and Szelenyi, M. (eds.), *Cities after socialism. Urban and regional conflict in post-socialist societies* (Oxford: Blackwell).

Socialism (2007), [homepage] Encyclopædia Britannica Online http://www.britannica.com/eb/article-9109587/socialism accessed on 13 June 2007.

Söderström, O. (2005), 'Representation' in Atkinson, D., Jackson, P., Sibley, D. and Washbourne, N. (eds), *Cultural geography. A critical dictionary of key concepts* (London: I.B. Tauris), 11–15.

Soja, E. (2001), 'Inside Exopolis. Scenes from Orange County' in Sorkin, M., *Variations on a Theme Park* (New York: Hill and Wang).

Sorin, A. and Tismaneanu, V. (eds) (2000), *Between past and future. The revolutions of 1989 and their aftermath* (Budapest: Central European University Press).

Śpiewak, P. (2005), *Pamięć po komunizmie* (Gdańsk: Słowo/obraz terytoria).

Stanilov, K. (2007), 'Taking stock of post-socialist urban development: A recapitulation' in Stanilov, K. (ed.), *The post-socialist city. Urban form and space transformations in Central and Eastern Europe* (Dordrecht: Springer).

Stanilov, K. (ed.) (2007), *The post-socialist city. Urban form and space transformations in Central and Eastern Europe* (Dordrecht: Springer).

Staniszkis, J. (2005), *Postkomunizm. Próba Opisu* (Gdańsk: Słowo/obraz terytoria).

Stankova, J. (1992), *Prague. Eleven centuries of architecture* (Prague: PAV).

Stenning, A. (2000), 'Placing (post-)socialism: the making and remaking of Nowa Huta, Poland'. *European Urban and Regional Studies*, 7/2, 99–118.

Stenning, A. (2005a), 'Out there and in here. Studying Eastern Europe in the West' *Area* 37(4), 378–383.

Stenning, A. (2005b), 'Post-socialism and the changing geographies of the everyday in Poland' *Transactions of the Institute of British Geographers* 30(1), 113–127.

Stevens, G. (1988), *The favored circle. The social foundations of architectural distinction* (Cambridge MA: MIT Press).

Sudjic, D. (1992), *The 100 mile city* (San Diego: Harcourt Brace & Company).

Sýkora, L. (1998), 'Commercial property development in Budapest, Prague and Warsaw' in Enyedi, G. (ed.), *Social change and urban restructuring in Central Europe* (Budapest: Akadémiai Kiadó).

Sýkora, L. (2007), 'Office development and post-communist city formation. The case of Prague' in Stanilov, K. (ed.), *The post-socialist city. Urban form and space transformations in Central and Eastern Europe* (Dordrecht: Springer).

Szarota, T. (ed.) (2001), *Komunizm. Ideologia, system, ludzie* (Warszawa: Wydawnictwo Neriton/Instytut Historii PAN).

Szcześniak, K. (2007), Wojewoda będzie walczył z Waryńskim. Dziennik Bałtycki 52(18947), 2–3.

Szczygieł, M. (2006), *Gottland* (Wołowiec: Czarne).

Szelenyi, I. (1996), 'Cities under socialism – and after' in Andrusz, G., Harloe, M. and Szelenyi, M. (eds), *Cities after socialism. Urban and regional conflict in post-socialist societies* (Oxford: Blackwell).

Szobor Park (2007), [homepage] http://www.szoborpark.hu/index.php?Lang=en accessed on 17 September 2007.

Szponar, A. (2003), *Fizjografia urbanistyczna* (Warszawa: Wydawnictwo Naukowe PNW).

Szyszkina, I. (1981), 'Współczesny okres rozwoju architektury radzieckiej: 1956–1980' *Architektura* 6(405), 19–23.

Tarachanow, A. and Kawtaradse, S. (1992), *Stalinistische Architektur* (Munich: Klinkhardt & Biermann).

Taverne, E. (2004), 'Rise and fall of the "second socialist city". Hoyerswerda-Neustadt' in Wagenaar, C. and Dings, M. (eds), *Ideals in concrete. Exploring Central and Eastern Europe* (Rotterdam: NAi Publishers).

Thomas, J. (1996), 'A precis of time, culture and identity' *Archeological Dialogues*, 1, 6–21.

Tilly, C. (1994), *A phenomenology of landscape. Places, paths and movements* (Oxford: Berg).

Traba, R. (2006), *Historia – przestrzeń dialogu* (Warszawa: Instytut Studiów Politycznych Polskiej Akademii Nauk).

Tuan, Y.-F. (1990), *Topophilia. A study of environmental perceptions, attitudes, and values* (Columbia University Press).

Tuan, Y.-F. (2005), *Space and place. The perspective of experience* (Minneapolis: University of Minnesota Press).

Turnbridge, J.E. and Ashworth, G.J. (1996), *Dissonant heritage: The management of the past as a resource in conflict* (Chichester: Wiley).

Turnbridge, J.E. (1998), 'The question of heritage in European cultural conflict' in Graham, B. (ed.), *Modern Europe. Place. Culture. Identity* (London: Arnold), 236–260.

Turner, V. (1975), *Dramas, fields, and metaphors. Symbolic action in human society (symbol, myth, & ritual)* (New York: Cornell University Press).

Unfield, B. (ed.) (1996), *Spuren des 'Realsozialismus' in Böhmen und Slowakei* (Wien: Löcker).

Universität Leipzig History (2006), 22 March 2006 [homepage] http://www.uni-leipzig.de/cumpraxi/english/history.html accessed on 20 September 2007.

Valentine, G. (2001), *Social geographies. Space & Society* (Harlow: Prentice Hall).

Wagenaar, C. (2004), 'Cities and the pursuit of public happiness. An introduction' in Wagenaar, C. (ed.), *Happy. Cities and public happiness in post-war Europe* (Rotterdam: NAi Publishers), 14–23.

Wagenaar, C. (ed.) (2004), *Happy. Cities and public happiness in post-war Europe* (Rotterdam: NAi Publishers).

Wagenaar, C. and Dings, M. (eds) (2004), *Ideals in concrete. Exploring Central and Eastern Europe* (Rotterdam: NAi Publishers).

Wagner, R. (2007), 'Im Zeichen der Halbwahrheit', *Neue Züricher Zeitung*, 31.08.

Waldheim, Ch. (2006), 'Landscape as urbanism' in Waldheim, Ch. (ed.), *The landscape urbanism reader* (New York: Princeton Architectural Press), 37–53.

Waldheim, Ch. (ed.) (2006), *The landscape urbanism reader* (New York: Princeton Architectural Press).

Wallach, B. (2005), *Understanding the cultural landscape* (New York: Guilford).

Wallraf, W. (2006), Urban Restructuring in the New Federal States, *Goethe-Institut Online*, [webpage] http://www.goethe.de/kue/arc/dos/dos/sls/wus/en1371040.htm accessed on 20 September 2007.

Wanklyn, H. (1961), *Friedrich Ratzel, a biographical memoir and bibliography* (Cambridge: Cambridge University Press).

Węcławowicz, G. (1997), 'The changing socio-spatial patterns in Polish cities' in Kovács, Z. and Wiessner, R. (eds), *Prozesse und Perspektiven der Stadtentwicklung in Ostmitteleuropa*, Münchener Geographische Hefte 76, Geographischen Institut der Technischen Universität München (Passau: L.I.S. Verlag), 75–81.

Węcławowicz, G. (1998), 'Social polarisation in postsocialist cities. Budapest, Prague and Warsaw' in Enyedi, G. (ed.), *Social change and urban restructuring in Central Europe* (Budapest: Akadémiai Kiadó).

Weiske, Ch. (2002), 'Stadt und Welt. Fiktive Verortungen als die Images der Stadt Chemnitz' in Hannemann, Ch., Kabisch, S. and Weiske, Ch. (eds), *Neue Länder – Neue Sitten? Transformationsprozesse in Städten und Regionen Ostdeutschlands* (Berlin: Schiler).

Wendell, S. (2003), *Stories I stole* (London: Atlantic Books).

Widgren, M. (2004), 'Can landscapes be read?' in Palang, H., Sooväli, H. and Antrop, M. (eds), *European rural landscapes. Persistence and change in a globalizing environment* (Boston: Kluwer Academic Publishers), 455–465.

Williams, R. (1982), *The Sociology of Culture* (New York: Shoken Books).

Winchester, H.P.M., Kong, L. and Dunn, K. (2003), *Landscapes. Ways of imagining the world* (Harlow: Prentice Hall).

Wódz, J. (1989), *Przestrzeń znacząca. Studia socjologiczne* (Katowice: Śląski Instytut Naukowy).

Wolff, R., Schneider, A., Schmid, Ch. and Klaus, Ph. (eds) (1998), *Possible urban worlds. Urban strategies at the end of the 20th century* (Basel: Birkhäuser Verlag).

Wołoszyn, W. (2006), 'Prawne i programowe podstawy ochrony krajobrazów kulturowych w Polsce' in Wołoszyn, W. (ed.), *Krajobraz kulturowy. Cechy, Walory, Ochrona*. Problemy Ekologii Krajobrazu, XVIII (Lublin: Zakład Ochrony Środowiska UMCS).

Wołoszyn, W. (ed.) (2006), *Krajobraz kulturowy. Cechy, Walory, Ochrona.* Problemy
    Ekologii Krajobrazu, XVIII (Lublin: Zakład Ochrony Środowiska UMCS).
Woźniczko, J. (2007), 'Petersburska bohema' *Przekrój,* 19/3229, 10.05, 44–48.
Zukin, S. (1993), *Landscapes of power. From Detroit to Disney World* (Berkeley:
    University of California Press).
Zukin, S. (1995), *The culture of cities* (Berkeley: University of California Press).
Zukin, S. (2000), 'Whose culture? Whose city?' in LeGates, R. and Stout, F. (eds),
    The city reader (London: Routledge).

Index